Interactive Sketch-Based Interfaces and Modelling for Design

RIVER PUBLISHERS SERIES IN DOCUMENT ENGINEERING

Series Editors:

STEVEN SIMSKE
Colorado State University, USA

Document engineering is an interdisciplinary set of processes and systems concerned with the analysis, design, development, evaluation, implementation, management, and/or use of documents and document corpora in order to improve their value to their users. In the era of the Internet, the millennia-old concept of a document is rapidly evolving due to the ease of document aggregation, editing, re-purposing, and reuse. In this series of books, we aim to provide the reader with a comprehensive understanding of the tools, technologies, and talents required to engineer modern documents.

Individual documents include web pages and the traditional print-based or print-inspired pages found in books, magazines, and pamphlets. Document corpora include sets of these documents, in addition to novel combinations and re-combinations of document elements such as mash-ups, linked sets of documents, summarizations, and search results. In our first set of books on document engineering, we will cover document structure, formatting, and layout; document structure; summarization; and classification. This set will provide the reader with the basis from which to go forward to more advanced applications of documents and corpora in subsequent years of the series.

The books are intended to cover a wide gamut of document engineering practices and principles, and as such will be suitable for senior undergraduate students when using the first 2/3 of each book (core topics), and be extensible to graduate students, researchers and professionals when the latter 1/3 of each book is also considered (advanced topics). Students and graduates of data analytics, information science, library science, data mining, and knowledge discovery will benefit from the book series.

For a list of other books in this series, visit www.riverpublishers.com

Interactive Sketch-Based Interfaces and Modelling for Design

Editors

Alexandra Bonnici
University of Malta, Malta

Kenneth P. Camilleri
University of Malta, Malta

River Publishers

Routledge
Taylor & Francis Group
NEW YORK AND LONDON

Published 2023 by River Publishers
River Publishers
Alsbjergvej 10, 9260 Gistrup, Denmark
www.riverpublishers.com

Distributed exclusively by Routledge
605 Third Avenue, New York, NY 10017, USA
4 Park Square, Milton Park, Abingdon, Oxon OX14 4RN

Interactive Sketch-Based Interfaces and Modelling for Design / by Alexandra Bonnici, Kenneth P. Camilleri.

Routledge is an imprint of the Taylor & Francis Group, an informa business

ISBN 978-87-7022-770-4 (print)
ISBN 978-10-0082-403-2 (online)
ISBN 978-1-003-36065-0 (ebook master)

While every effort is made to provide dependable information, the publisher, authors, and editors cannot be held responsible for any errors or omissions.

Contents

Preface

This book is part of the River Series in Document Engineering. As part of this series, we explore the use of sketches created during the design process, looking at the tools and algorithms that facilitate their use in engineering design.

Document engineering is the science that focuses on the design and implementation of systems centered around documents and document collections. Thus, document engineering presents techniques that enable the creation, management, sharing, and productive use of documents.

The document engineering approach is not restricted to text-based documents but applies to any domain where documents are generated. In design engineering, these documents consist primarily of 2D and 3D sketches, virtual prototypes and models, all of which support the final tangible product or artwork. In this domain, document authoring traditionally involved pen-and-paper sketching. However, the increasing availability of computing technologies has seen a change in authoring tools, shifting from paper to digitizing tablets, and now, to virtual and augmented reality environments. In this book, particularly in Chapters 6–8, we examine authoring tools and techniques that enable 3D sketching in space.

Despite new technologies, traditional pen-and-paper sketching remains prevalent. Thus, just as optical character recognition is required to convert hand-written documents into digital documents, so is the need for sketch recognition techniques that support the automated generation of 3D models. Chapters 2–4 explore sketch simplification and interpretation algorithms for both physical and digital paper.

Document search interactions that allow retrieving of other relevant documents have a similar role in the design process. Here, rather than text-based search, the search query would be a sketch and the searchable material 3D renders or images of physical objects. The difficulty here lies in reconciling the sketch with its inherent ambiguities and the realistic-looking counterparts. The study of non-photorealistic rendering techniques builds an understanding of how people sketch and what salient features people choose to sketch, and thus, how to formulate sketch-based queries. Chapter 5 of this book describes these concepts.

This book, therefore, brings together the sketch interaction techniques that address the challenges in these areas, opening the domain of sketch understanding and engineering design to other techniques from the document engineering domain. Specifically, techniques from document version control may be introduced to the engineering design pipeline to record design decisions are taken and their impact on the final product design.

List of Acronyms

2D two-dimensions

3D three-dimensions

ADG angular distribution graph

API application programming interface

AR Augmented reality

ASL American sign language

CAD Computer-aided design

CAM computer-aided manufacturing

CAVE cave automatic virtual environment

CBIR content based image retrieval

CGAN conditional generative adversarial network

CNN convolutional neural network

CPU central processing unit

CRT cathode-ray tube

CT computed tomography

DoF degrees of freedom

DoG difference of Gaussians

DRIVE digital retinal images for vessel extraction

DSIFT dense scale invariant feature transform

EMS electrical muscle stimulation

FoV field of view

GA genetic algorithm

GALIF Gabor local line-based feature

GAN generative adversarial network

GLM generalised linear model

globalPb global probability of boundary

GPU graphics processing unit

GUI graphical user interface

HCI Human-computer interaction

HMD head-mounted display

HMM hidden Markov model

HOG histogram of oriented gradients

IMU inertial measurement unit

IR infrared

LCD liquid crystal display

LED light-emitting diode

LSTM long short-term memory

MFAT multiscale-fractional anisotropic tensor

MRI magnetic resonance imaging

NPR non-photorealistic rendering

OLED organic light-emitting diode

ReLU rectified linear unit

ROC receiver operating characteristic

SIFT scale invariant feature transform

SkRUI sketch recognition user interfaces

SPE salient point error

SVD singular value decomposition

SVM support vector machine

UI user interface

VR virtual reality

WIMP windows, icons, menus, pointer

XR extended reality

List of Figures

List of Contributors

Arora, Rahul, *University of Toronto, Toronto, Canada*

Bonnici, Alexandra, *University of Malta, Malta*

Camilleri, Kenneth, *University of Malta, Malta*

Israel, Johann Habakuk, *Hochschule für Technik und Wirtschaft Berlin, University of Applied Sciences, Germany*

Keefe, Daniel *University of Minnesota, US*

Liu, Juncheng, *University of Otago, New Zealand*

Machuca, Mayra Donaji Barrera, *Dalhousie University, Canada*

Metin Sezgin, T., *Koç University, Turkey*

Rosin, Paul L., *Cardiff University, Wales, UK*

Wacker, Philipp, *RWTH Aachen University, Germany*

1

Introduction

Alexandra Bonnici and Kenneth P. Camilleri

University of Malta, Malta

The first examples of art known to man date to the Cro-magnon people. The artistic artifacts from this period are beautiful and seem to capture the movement of the subjects portrayed. They are an art form which indicates that from the earliest of times man sought to represent not just the world around him but also the spirit of the subject he portrays [20, chap. 1]. Today, despite easy access to computer aids, drawing and sketching remain relevant as tools to communicate ideas and form concepts. The versatility of drawings allows their use across a variety of applications, from illustrations of complex anatomical structures to instruction manuals, from architectural plans to route maps. This versatility becomes more evident when one takes into consideration that drawings are used across cultures and across times [22, 16].

In this book, we will focus on the use of sketching in design and product manufacturing, and in particular, in the use of sketches in human-computer interaction systems to aid the design process. Modern demands on the manufacturing industry require more consumer-centric designs than ever before. Designers are expected to produce innovative products and to meet increasing requirements for customization. Thus designers face lower volume requests and higher product-mix to meet the individual consumer needs while at the same time needing to retain low costs to remain competitive. Trends for globalization in manufacturing as well as growing environmental concerns, lead to more complex manufacturing processes. The industry, therefore, requires systems that support communication between the different stages of the product development life-cycle and which help designers reach decisions faster and with fewer errors [13].

Moreover, in recent years, the accessibility of 3D printers and 3D printing services have also changed the way products are fabricated. This increased accessibility shifted the fabrication process from the traditional factory environment to a more diverse user-base, fabricating a wider range of products, from jewelry to furniture and more [15]. The diversity of the user-base brings with it users that have different skills and expertise levels, and this too raises the need for tools and systems that facilitate the transition from concept ideas to physical products.

Computer-aided design (CAD) and computer-aided manufacturing (CAM) systems are increasing the support offered to the product design and manufacturing industries, including through the integration of Augmented reality (AR) and virtual reality (VR) systems, which are collectively known as extended reality (XR) systems. Image and texture analysis and object recognition allows the projection of digital images and textures onto real objects or augmentation of the real world with virtual objects. Likewise, techniques such as haptic feedback and object tracking, provide tangible support for grasping and manipulating objects. With this support, designers can create virtual prototypes of their designs to visualize the object as a three-dimensional form, inspect and interact with the object as well as simulate its functionality without the costs of creating physical prototypes [16]. Such XR systems also facilitate collaborative tasks, allowing for both on-site and off-site or remote collaborations.

Despite the advantages and widespread dissemination of CAD and CAM systems, designers still rely on traditional pen-and-paper sketching. One reason for this is that in contrast with pen-and-paper sketching, computer-supported design systems tend to be cumbersome and quite rigid [25]. Some CAD systems, with complex user interface, divert the designer's attention from the design onto the user interface. Other CAD systems require design decisions such as exact dimensions, or material properties, too early in the design process, such that reliance on CAD systems reduces the abstraction capacities of the design [4]. Thus, CAD tools may, at times, hinder rather than aid the early design process. The positive aspects of CAD, that is, the ability to visualize and interact with 3D forms, are, however, desirable. Hence, the challenge in the development of these CAD tools lies in the creation as interfaces that offer the same simplicity, portability, flexibility, and fluidity of the traditional pen-and-paper tools [25].

To address the challenge of suitable interfaces, we need first to establish the reasons why designers sketch and what makes sketching important in engineering design.

1.1 Why Do People Sketch?

In design-oriented disciplines such as architecture and engineering as well as in visual communication disciplines, drawings are tools of creativity and problem solving [3]. Designers will typically create different drawings throughout the design life-cycle. These drawings will differ in style and formality to reflect the different requirements and needs of the design process. In the initial stages of the design, the designers use *concept sketches*, a collection of visual cues that are sufficient to suggest a design to an informed observer. In this sketching stage, the designer is typically exploring ideas and possibilities without committing to a particular design or approach. We may further describe the concept sketch in terms of the physical elements that are present in the sketch [26]. Thus, a *level 1* sketch would be a simple, monochromatic line drawing with no shadows or annotations. A *level 2* sketch would have additional annotations while a *level 3* sketch introduces basic shadows to suggest 3D form. A *level 4* sketch uses shading gradations and color to place further emphasis on the 3D form.

The *presentation drawing* is often drawn from the concept sketch. Through these drawings, the designer can present to a client a selection of highly rendered drawings that are as realistic as possible to help in the formation of the design decisions. After establishing the design, the designer, or other engineers/technicians will draw the *general arrangement drawings*. These drawings focus on the functionality and inner workings of more complex systems. At this stage of the process, the designer will add the dimensions required for assembly and production. Finally, after finalizing all design decisions, *technical illustrations* are produced. These drawings provide the necessary information for producing, assembling, and maintaining the product [14, Ch. 1].

While not all designs will require the creation of all the drawings outlined above, almost all designs start with a concept sketch. The concept sketch is, therefore, considered an essential part of the design process. It acts as a cognitive tool that can be used to augment memory and for information processing, relieving the mind from the burden of simultaneously holding content in memory while operating on it. Sketches are also a means of recording information and communicating this information to others and oneself. Thus, the sketch has a similar role to speech. However, the sketch can also communicate visuospatial ideas directly; hence, the advantage of sketching design ideas rather than describing them verbally. In the design process, the sketch helps the designer check for completeness of the idea, and explore new relations which may stem from this initial idea [22].

These observations stem from experimental observations in various design disciplines. For example, Schütze et al. conducted an experiment with industrial designers in which, designers were asked to solve specific design problems with specific constraints. Schütze et al. found that the designers who are allowed to sketch are more likely to provide solutions with better functional quality and have fewer difficulties in the design process than those who are not [18]. Likewise, sketching has been shown to provide more support for individual re-interpretation cycles, enhancing individual and group access to earlier ideas [24]. Earlier experiments in interface design also demonstrate that sketching ideas helps design teams focus on broader conceptual ideas. In this manner, designers are not distracted by relatively minor issues such as fonts or the alignment of objects [11].

1.2 Why Are Concept Sketches Conducive to Design Creativity?

Unlike other, later-stage drawings, concept sketches will typically appear "sketchy", that is, they appear vague and non-committed, pinning down only the minimal global arrangements and figures of the design concept. While this may make the sketch ambiguous, confusing, or uncertain, this lack of defining features gives the sketch more vitality and strength than well-defined sketches [20, chap. 2]. The incomplete and ambiguous nature of the sketch encourages designers to focus on different parts of the sketch, mentally rearranging parts of the sketch to make inferences about the design and see new interpretations. Designers become adept at making new inferences from their design sketches or the sketches of others. These inferences may be functional, for example, seeing the flow of pedestrians in a sketched floor plan, or perceptual, for example, seeing new spatial relations among structures [23]. Thus, the ambiguity and incomplete nature of the sketch help the designer from fixating on old ideas and keep generating new ones. By removing constraints gives the designer free reign to think "outside the box" without the need to consider physical laws or the manufacturability of the design and this is conducive to creativity.

1.3 Is Ambiguity in Sketches Always Desirable?

From an interpretation point of view, a drawing is a series of line marks that form some symbols on the paper [7]. Lines in the drawing have a definite shape and size which are suggestive of proportions and magnitudes. Thus,

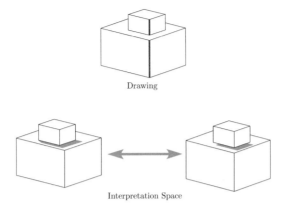

Drawing

Interpretation Space

Figure 1.1   A drawing can have a range of interpretations assigned to it by different viewers. In this example, the alignment of the two vertical lines shown in red allows interpretations which place the smaller box centered on top of the larger box to others which place the smaller box floating on top of the larger box.

each line has a *perceptual interpretation space* which defines the range of meanings that the line may take.

For a viewer to understand and interpret the drawing, the viewer needs to perceive both the symbolic categories and the shape of the design elements. Thus, the drawing also has a *deductive interpretation space*, which defines the range of interpretations assigned to the drawing by the viewer [19] as illustrated in Figure 1.1. The centers of these interpretation spaces define the interpretation most strongly suggested by the drawing. Thus, for the sketcher and the viewer to agree on an interpretation, the perceptual and deductive interpretation spaces must overlap.

However, when people sketch quickly, the accuracy of their sketching diminishes, leading to the roughness or sloppiness associated with sketches. This roughness increases the width of the interpretation space, giving rise to the varying interpretations of the sketch. Different people make different judgments on what qualifies as imprecision and roughness, such that the width of the interpretation space varies from viewer to viewer, varying the tolerance to alternative interpretations. Moreover, designer idiosyncrasies and even poor drawing styles of the sketcher introduce a bias to the interpretation space. This bias further affects the interpretation of the sketch by the different people involved in the design process, as illustrated in Figure 1.2. Exploring these multiple interpretations of the sketch at the early concept-design stage of the product life-cycle may result in the revelation of new insights

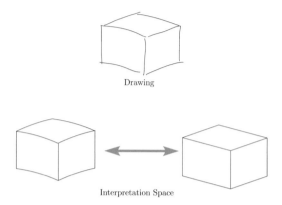

Drawing

Interpretation Space

Figure 1.2   In quick, sloppy sketches, the viewer's tolerance to roughness changes the interpretation of the sketch. In this simple example, varying tolerance to roughness will change the interpretation of the sketch from one with curved sides to one with straight sides.

and facilitates the ad-hoc discovery of new interpretations and is therefore desirable in the sketching process [5, 21].

   If we further consider the sketch as the representation of a 3D shape, we have the added complication of the sketch being a projection of a 3D object onto a flat, paper medium. In theory, any number of 3D shapes can project onto the same sketch [12]. Despite the numerous possible interpretations, the sketch is likely to have a strong perceptual center. Such a strong perceptual center arises because, through experiences, we discard alternative projections as highly improbable [7, 12]. The Gestalt laws of perceptual grouping allow us to discern between likely and unlikely closed paths [17], further assisting the selection of possible face contours. However, multiple interpretations of sketched 3D shapes are indeed possible as illustrated by the Necker cube or Schroeder's staircase illustrations shown in Figure 1.3. While these pictures are reminiscent of Escher's "Impossible Constructions"[1] and are unlikely to occur in real concept sketches, similar ambiguities may arise in badly or sloppily drawn sketches. Such sketches may exhibit regions that are locally consistent, but the overall sketch is globally inconsistent. As a result, focusing on different parts of the sketch leads to different interpretations [7]. In this case, the perception center shifts from one interpretation to another with the strongest interpretation being dependent on the user bias. Such ambiguities would require clarification before further processing of the sketch occurs.

---

[1]https://mcescher.com/gallery/impossible-constructions/

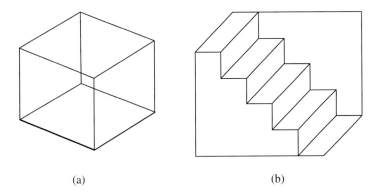

(a)   (b)

Figure 1.3 Drawings for which multiple interpretations can co-exist. Figure (a) shows the Necker cube while Figure (b) shows Schroeder's staircase. In both examples, the interpretation space includes strong support for different interpretations and the interpretation preferred by the viewer depends on their natural biases.

While the ambiguities mentioned above relate to the interpretation of the sketch as a drawing artifact, ambiguities may also arise when designers use the sketch as a communication tool in a collaborative environment. Two scenarios are possible. In the first scenario, designers in collaboration are interacting with the sketch simultaneously. In this scenario, there are two possible sources of ambiguity. The first arises when the focus of attention of the participating designers is different from one another and, as a result, the designers will unintentionally communicate about different parts of the sketch. The second source of ambiguity occurs when designers use symbols with differing scopes within the same collaborative environment [5].

The second scenario occurs when designers and engineers work in different stages of the product life-cycle. Although we would assume that proper communication or hand-over between designers, engineers, and technicians takes place, this is not necessarily always the case [19]. In the absence of clear communication of intent, the drawing ambiguities discussed above resurface further down the design cycle, potentially leading to bad design decisions.

In both scenarios, communication through the sketch would need to be complete and unambiguous to ensure that the design is a result of the shared understanding, which reflects the original intent and which meets the concerns of all [19]. Thus while ambiguity arising from the rough nature of the sketch encourages design discovery and is desirable, ambiguity through miscommunication may hinder the design process and should be discouraged.

1.4 How Can Computer Systems Address These Problems?

Computer systems can combine processing power with the nonrestrictive nature of sketches to allow designers to perceive and evaluate product properties under their multimodal and 3D forms. To achieve this potential requires combining efforts from several research areas including computer graphics, machine learning, augmented and virtual reality environments, tracking, and image/object retrieval, among others. An effective CAD system would be one that allows for a natural interface that does not distract the user from the design process. It should strike the right balance between creating a representation that allows for realistic rendering and incompleteness to allow for functional simulation while at the same time, leaving space for design thinking [16]. The CAD system should support sketching using the designer-preferred drawing medium. This medium could be the traditional pen-and-paper, thereby requiring sketch interpretation systems, or the 3D space itself, requiring the ergonomic support necessary to sketch in space. Moreover, the CAD system should provide the tools necessary for unambiguous communication among multi-person collaborators, providing the means for assessing and re-directing the focus of attention of multiple users.

Research communities are tackling these issues in various ways. For example, Ku et al. [10] apply line tracking algorithms to determine the sketched line strokes and use Gestalt-like principles to group these line strokes into meaningful groups, while Cao et al. [1] apply 3D geometry to create 3D models from paper-based drawings, thus, interpreting flat, paper-based drawings as 3D forms, not unlike the human interpretation of the drawing. This interpretation, is, however, limited by what is drawn by the user, deducing the unsketched 'back' of the object through heuristic rules based on human perception. On the other hand, sketch-based interfaces such as Teddy [8], among others, follow an interactive approach where the user can manipulate the object through sketched gestures. Such an interaction gives the user more control over the way the object looks and allows for progressive manipulation of the object from different viewpoints. Despite creating 3D objects, the interactions with these interfaces remain through a

3D medium. Alternative interactions such as that proposed in Keefe et al. [9] use tangible user interfaces to allow the user to sketch directly in the 3D space. Here, tracking of gaze and head rotations allows the determination of the point of gaze and this, can then, in turn, be used to guide the focus of attention in collaborative environments [6]. non-photorealistic rendering (NPR) can be used to reach the desired balance between a realistic render of the object and the incompleteness of the design. Thus, observing how people depict objects as sketches allows researchers like Cole et al. [2] to render objects using sketch-like strokes, retaining the underlying accuracy of the object shape while displaying a style that suggests incompleteness.

Thus, there are different research tools available that address specific problems in the early concept design stage, helping to bridge the gap between early concept sketching and the later stages in the design process. The integration of these systems in product design helps to make the design process accessible to non-expert users. This helps with the customization of products, allowing the client to play an integral part in the design process. By quickly and effortlessly changing a 3D sketch into a 3D virtual prototype through which the client and the designer may explore functionality issues, will also allow the designer to explain design decisions to clients, making the design process more transparent.

1.5 Organisation of the Book

In this book, we take a broad view and look at the interpretation problem in diverse contexts for example, in the context of 3D modeling, sketch-based retrieval, multimodal interaction, virtual and augmented reality interfaces.The rest of the book is divided into two parts. The first part treats the use of 3D sketches and drawings, focusing on offline drawing simplification and vectorization in Chapter 2, online drawing simplification and recognition in Chapter 3, the interpretation of drawings as 3D objects in Chapter 4 and sketch-based search interactions in Chapter 5. The second part of the book treats sketching in the XR environments. This part introduces the origins of 3D sketching in Chapter 6, describing the input processing techniques and geometric representations for 3D sketches in Chapter 7, the different interaction devices and techniques for 3D sketching in Chapter 8, concluding this part with a description of different application scenarios in Chapter 9. Chapter 10 concludes the book by discussing open questions and future directions.

References

[1] L. Cao, J. Liu, and X. Tang. What the Back of the Object Looks Like: 3D Reconstruction from Line Drawings without Hidden Lines. *IEEE Transactions on Pattern Analysis and Machine Intelligence*, 30 (3):507–517, 2008.

[2] F. Cole, A. Golovinskiy, A. Limpaecher, H. S. Barros, A. Finkelstein, T. Funkhouser, and S. Rusinkiewicz. Where Do People Draw Lines? *ACM Transactions on Graphics*, 27(3):1–1, 2008.

[3] B. Craft and P. Cairns. Sketching Sketching: Outlines of a Collaborative Design Method. In *Proceedings of the 23rd British HCI Group Annual Conference on People and Computers: Celebrating People and Technology*, BCS-HCI '09, pages 65–72.

[4] T. Dorta, E. Pérez, and A. Lesage. The Ideation Gap: Hybrid Tools, Design Flow And Practice. *Design Studies*, 29(2):121–141, 2008.

[5] M. J. Eppler, J. Mengis, and S. Bresciani. Seven Types Of Visual Ambiguity: On The Merits And Risks Of Multiple Interpretations Of Collaborative Visualizations. In *12th International Conference on Information Visualisation*, pages 391–396.

[6] X. He and Z. Liu. A Novel Way Of Estimating A User's Focus Of Attention In A Virtual Environment. In Jessie Y. C. Chen and Gino Fragomeni, editors, *Virtual, Augmented and Mixed Reality: Interaction, Navigation, Visualization, Embodiment, and Simulation*, pages 71–81, Cham, 2018. Springer International Publishing.

[7] D. D. Hoffman. *Visual intelligence: How we create what we see.* WW Norton & Company, 2000.

[8] T. Igarashi, S. Matsuoka, and H. Tanaka. Teddy: A Sketching Interface for 3D Freeform Design. In *Proceedings of the 26th Annual Conference on Computer Graphics and Interactive Techniques*, SIGGRAPH '99, pages 409–416, 1999.

[9] D. F. Keefe, D. A. Feliz, T. Moscovich, D. H. Laidlaw, and J. J. LaViola. CavePainting: A Fully Immersive 3D Artistic Medium and Interactive Experience. In *Proceedings of the 2001 Symposium on Interactive 3D Graphics*, I3D '01, pages 85–93, 2001.

[10] D. C. Ku, S. F. Qin, and D. K. Wright. Interpretation Of Overtracing Freehand Sketching For Geometric Shapes. In *14th International Conference on Computer Graphics, Visualization and Computer Vision*, pages 263–270. Václav Skala-UNION Agency, 2006.

[11] J. A. Landay and B. A. Myers. Interactive Sketching for the Early Stages of User Interface Design. In *Proceedings of the SIGCHI Conference on Human Factors in Computing Systems*, CHI '95, pages 43–50, 1995.

[12] H. Lipson and M. Shpitalni. Conceptual Design and Analysis by Sketching. *Artificial Intelligence for Engineering Design, Analysis and Manufacturing*, 14(5):391–401, November 2000.

[13] A. Y. C. Nee, S. K. Ong, G. Chryssolouris, and D. Mourtzis. Augmented Reality Applications in Design and Manufacturing. *CIRP Annals*, 61(2): 657–679, 2012.

[14] A. Pipes. *Drawing for Designers*. Laurence King Publishing, 2007.

[15] T. Rayna and L. Striukova. From Rapid Prototyping To Home Fabrication: How 3D Printing Is Changing Business Model Innovation. *Technological Forecasting and Social Change*, 102:214–224, 2016.

[16] V. Rieuf, C. Bouchard, V. Meyrueis, and J. F. Omhover. Emotional Activity In Early Immersive Design: Sketches And Moodboards In Virtual Reality. *Design Studies*, 48:43–75, 2017.

[17] E. Saund. Finding perceptually closed paths in sketches and drawings. *IEEE Transactions on Pattern Analysis and Machine Intelligence*, 25(4): 475–491, 2003.

[18] M. Schütze, P. Sachse, and A. Roemer. Support Value of Sketching in the Design Process. *Research in Engineering Design*, 14:89–97, 2003.

[19] M. Stacey and C. Eckert. Against Ambiguity. *Computer Supported Cooperative Work*, 12(2):153–183, June 2003.

[20] F. Thomas and O. Johnston. *The Illusion of Life: Disney Animation*. Hyperion, New York, 1995. ISBN 9780786860708.

[21] W. S. W. Tseng and L. J. Ball. How Uncertainty Helps Sketch Interpretation In A Design Task. In *Design Creativity 2010*, pages 257–264. Springer, 2011.

[22] B. Tversky. What Do Sketches Say About Thinking. In *AAAI Spring Symposium, Sketch Understanding Workshop, Standford University, SS-02-08*, pages 148–151, 2002.

[23] B. Tversky, M. Suwa, M. Agrawala, J. Heiser, C. Stolte, P. Hanrahan, D. Phan, J. Klingner, M. P. Daniel, P. Lee, and J. Haymaker. *Sketches For Design And Design Of Sketches*, chapter 7, pages 79–86. Springer Berlin Heidelberg, Berlin, Heidelberg, 2003. ISBN 978-3-662-07811-2.

[24] R. van der Lugt. Functions of Sketching in Design Idea Generation Meetings. In *Proceedings of the 4th Conference on Creativity & Cognition*, C&C '02, pages 72–79, 2002.

[25] M. Xin, E. Sharlin, and M. Costa Sousa. Napkin Sketch: Handheld Mixed Reality 3D Sketching. In *Proceedings of the 2008 ACM Symposium on Virtual Reality Software and Technology*, VRST '08, pages 223–226, 2008.

[26] M. C. Yang. Observations on Concept Generation and Sketching in Engineering Design. *Research in Engineering Design*, 20(1):1–11, March 2009. ISSN 1435-6066.

Part I

Sketches Drawn on 2D Media

2

Simplification and Vectorization of Sketched Drawings

Alexandra Bonnici and Kenneth P. Camilleri

University of Malta, Malta

The need to process and extract information from drawings stems from practitioners in the fields of engineering, architectural design, and cartography among others, with the need to process paper-based drawings. These drawings need to be digitized as raster images from which all line strokes are then extracted. All processing of the drawing is carried out *after* drawing completion, hence the term offline processing. Although the widespread availability of digital tablets saw a shift towards digital-born drawings, offline processing of drawings remains relevant since the paper medium remains in use by artists and designers [55]. Moreover, digital-born drawings may be saved in either raster format or as time-stamped vectors. Digital-born drawings saved as raster images require the same processing as their paper-based counterparts. Thus, the difference between online and offline sketch processing lies not in the medium used, but in the stroke information available which then affects the possible methods of interpretation. In offline sketching applications, the user must complete the drawing before this is interpreted as a whole in a mostly noninteractive manner. This chapter will focus on techniques required for offline sketch processing and simplification.

The first step towards the interpretation of offline drawings involves the extraction of line strokes that define the edge boundary of the object from the sketch, a process known as drawing vectorization. For neat and simple drawings, vectorization can follow three basic steps, namely,

Image binarization which separates the sketched strokes from the image background,

15

Line localization that divides the drawing into individual, pixel-wide strokes,

Curve fitting which fits geometric curves and lines to the thinned strokes, thereby representing the object edges as neat vectors.

An overview of binarization algorithm performance of traditional and emerging binarization algorithms under different light conditions can be found in [26]. Here, we will focus on the subsequent two steps of the vectorization problem, that is, the line localization and the curve fitting problems.

2.1 Line Localization

The line localization step is required to remove the redundant black pixels which are a result of the thickness of the pen/pencil tool with which the drawing is made and to retain only the pixels which define the geometry of the sketched object. While skeletonization is a fast, easy and efficient way to obtain the desired single-pixel wide lines, off-the-shelf skeletonization techniques are known to displace the localization of intersection points between two or more lines and also introduce short, spurious line-segments towards the end-points of line strokes as shown in Figure 2.1(a) [25]. These displacements and spurs become more evident with thicker lines, or at intersection regions and result in distortions of the final object representation [17].

Corrective measures based on some heuristics may be applied at the curve fitting step to correct for these distortions. Alternatively, the line localization procedure can be adjusted to obtain a more accurate representation of the line strokes. The latter approach becomes more important if drawing interpretation is to allow more flexibility in drawing styles. Thus, Chiang [8] proposes to use the maximum inscribed circle to thin the strokes into single-pixel wide lines, thereby avoiding spurs at line endpoints. Janssen and Vossepoel [20] use off-the-shelf skeletonization, but the skeleton is used to find the approximate location of the junction points. They then use idealized junction shapes and fine-tune the position of the junction through morphological operations with the idealized junctions. Hilaire and Tombre [19] are driven by similar idealized junctions, but use junction-vertex graphs to determine the topology of the drawing and hence adjust the junction position.

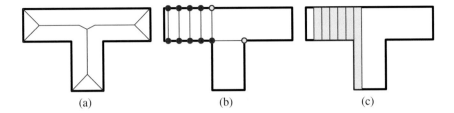

Figure 2.1 Examples of line localization techniques: (a) skeletonization introduces spurs and displaces junction points; (b) contour matching methods identify junction regions from the one-to-many mappings between contour pixels; (c) run-length approaches identify junction regions from irregular run-lengths.

It is also possible to use the stroke contours for line localization The concept here is that at non-junction points, each contour pixel will have an opposite contour point and there is a one-to-one mapping between pairs of opposite contour points. At junctions, however, the one-to-one mapping breaks down to a one-to-many mapping as shown in Figure 2.1(b), hence indicating the onset of a junction region. Pairs of contour points can be modeled by straight lines [22, 16] or trapezoids [37] such that the junction position may be extrapolated from the more regular line regions. The obvious drawback of using the stroke contours is that it becomes necessary to match opposite pairs of contour points for line segments which are fairly well represented by single skeletal lines. Thus, Tanigawa et al. [49] and Röösli and Monagan [38] use both the skeleton and the line contours for vectorization. Here, the stroke's skeleton is used for regular line segments, identifying junction regions through the valence of the skeletal points. At junction regions, the algorithm reverts back to the stroke contours for the correct positioning of the junction point.

Following similar logic, Boatto et al. [5], Monagan and Roosli [31], Di Zenzo et al. [12] and Keysers and Breuel [23] represent the line strokes using run-lengths as shown in Figure 2.1(c). Run-length scans have also been used to segment surfaces of digitally rendered range images into planar regions [21]. Within the context of vectorization, the reasoning is that regular lines can be represented by horizontal or vertical run-lengths with near-equal lengths such that the stroke center-line can be obtained from the midpoint of the run. At junction points, however, the length of the run changes, and this identifies the junction region. Perturbations on the line stroke, whether due to digitization or varying stroke thickness, may affect the length and regularity of the runs such that line modeling is also required to ensure the robustness of the line and junction localization.

Line localization can also be obtained through other image processing techniques such as the Hough transform, which accumulates votes from pixels belonging to the same line parameters in the accumulator array. The task of the line localization is therefore that of finding the peaks in the accumulator array and locating the endpoints of the lines with the corresponding parameters in the image space. Difficulties may arise due to the stroke widths and additional spurious line strokes which would diffuse the locality of the peak in the accumulator array. So, vectorization algorithms that use the Hough transform constrain the voting in the parameter space. For example, Olson [34] and Song and Lyu [47] use the line width estimation to aggregate pixels in the image space such that voting in the accumulator space occurs by pixel groups rather than by individual pixels, while Guerreiro and Aguiar [18] use the Hough transform in conjunction with stroke contours, using the edge connectivity and orientation as constraints on the parameter voting.

The methods described thus far require the processing of each individual pixel forming the line stroke to determine its candidacy as a skeletal or contour point, while all the resulting skeletal/contour points are again assessed for their contribution to the line/curve model. For the greater part of the line strokes, such a fine sampling of the strokes is unnecessary since the strokes exhibit stable characteristics. Finer sampling becomes necessary as the line approaches junction points or endpoints. This observation brought about sparse-pixel tracking methods which sparsely sample the drawing using rectangular segments [14, 48] or squares [15, 32], starting from an initial seed segment in a suitable place in the image. The interaction between the sampling segment and the line drawing is then used to estimate the direction of propagation of the segment. The size of the segment will, therefore, determine the sparsity of the sampling. Since the ink strokes are expected to have a rectilinear shape, the rectangular samplers have the advantage of having a similar shape as the lines being sampled [14]. However, the rectangular segments must be aligned with the lines being extracted. This will involve measuring the degree of overlap at each sample point, rotating the sampling segment and adjusting its width in order to maximize the match between the sampler and the line drawing [14, 48].

Rather than using a rectangular segment, El-Harby and Meyer [15] and Nidelea and Alexei [32] use a square sampler, noting the number of black pixels on the image border in order to detect the tracking direction.

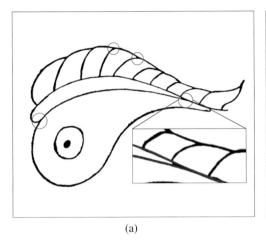 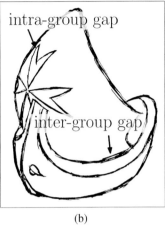

(a)   (b)

Figure 2.2 (a) Free-form shapes have lines which intersect at more acute angles (examples circled). (b) Concept sketches are likely to contain over-sketched strokes.

Thus, rather than adjusting the orientation and position of the sampler to the line stroke such that a linear path propagation can be employed, these algorithms adjust the tracking direction and in so doing, align the sampler to the line strokes. However, the square sampling segment is not isomorphic such that the intersection between the sampler and the lines will be different depending on the orientation of the line with respect to the sampler. A better sampling approach is adopted in [7, 33, 54] whereby circular samplers are used for vectorization, allowing for a more consistent sampling of the edges, particularly around junctions.

The algorithms described above were developed principally with the scope of vectorizing engineering or technical drawings, architectural drawings, or cadastral maps. The drawings expected by these algorithms, therefore, consist mainly of representations of rigid objects with the drawings potentially made with the aid of drawing tools, such as rulers, therefore resulting in neat, two-toned drawings where line strokes can be easily distinguished from the paper background. More recently, researchers are shifting the focus to freehand drawings of objects that exhibit free-form or organic shapes, that is shapes consisting of a flowing and curving appearance. This introduces three problems which well-established vectorization algorithms do not address. First, as illustrated in Figure 2.2(a), organic shapes will have instances where line strokes appear to merge into

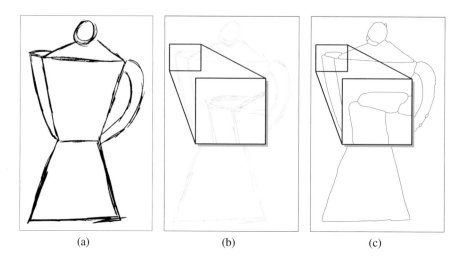

(a)   (b)   (c)

Figure 2.3   (a) Example of an image with oversketching. (b) Edge detection will detect the edges of each line stroke resulting in redundant edges. (c) Applying morphological closing followed by skeletonization. Here the closing operation bridges some of the gaps between oversketched line strokes, but exacerbates the displacement of the junctions

each other or intersect at sharper angles than what is typical in drawings of rigid objects. As a result, it becomes more critical to obtain the proper resolution of the line strokes as they approach the junction region. Secondly, the drawings are more likely to contain over-sketched strokes, that is, instances where the object boundary is represented by multiple ink strokes rather than one single, continuous line as shown in Figure 2.2(b). The third issue relates to shading that may be present in the image and which adds to the perception of depth in the drawing.

Since most vectorization algorithms assume that the image can be correctly binarized, the first problem is encountered at this pre-processing step. The shading causes the image histogram to no longer remain bimodal and thus, binarization would typically cause misclassification at the shaded regions. Moreover, edge detection algorithms are likely to find false edges at the shading-to-background boundary, or indeed, at the line strokes that form the shading itself, thereby distorting the geometry of the object. Putting the shading problem aside, skeletonization as well as edge detection algorithms will perform poorly when there are over-sketched line strokes present in the drawing, as shown in Figure 2.3(b), capturing the individual strokes rather than treating these as one single line stroke. Although it is possible to

Box 2.1 Binary Morphology

In binary morphology, a binary image is compared to a probing shape or *structuring element*, to establish how this fits within the foreground shape. Depending on the shape of the structuring element and the binary morphology operation used, the foreground shape will be altered. Consider a binary image A and a structuring element B. The basic binary morphology operations are defined as described here-under.

Erosion is defined as $A \oplus B = \{x | B_x \subseteq A\}$ where B_x is the translation of B by the vector x. Erosion has the effect of removing from a shape in A the outer layer according to the shape and size of B, effectively making the shape appear skinnier. In sketch applications, erosion removes the thickness of line strokes.

Dilation is defined as $A \ominus B = \{x | \hat{B}_x \cap A \neq \Phi\}$ where $\hat{B}$ is the reflection of B and Φ is the empty set. Dilation has the effect of adding to a shape in A an outer layer according to the shape and size of B, effectively making the shape appear thicker. In sketch applications, dilation adds thickness to the sketched lines.

Opening is obtained by eroding the image A by B then dilating the result back by B that is, $A \circ B = (A \ominus B) \oplus B$. In sketch applications, this would have the effect of removing digitization artifacts from the line strokes, resulting in a smoother stroke contours without changing the line thickness.

Closing is obtained by dilating the image A by B then eroding the result back by B, that is, $A \bullet B = (A \oplus B) \ominus B$. In sketch applications, this would have the effect of closing gaps between over-sketched strokes.

close gaps between the over-sketched strokes through morphology operations (see Box 2.1) as shown in Figure 2.3(c), such operations may exacerbate the problems with the inappropriate resolution of line strokes and junction locations. In the following, we give an overview of how these problems are being addressed.

2.1.1 Vectorization of Over-Sketched Drawings

To understand the sketch simplification problem, it is worth understanding better the characteristics of the rough sketches. As shown in Figure 2.2(b) the rough sketch consists of strokes that are inter-spaced by gaps. These strokes will be drawn in such a manner that they are loosely aligned with the edges of the object that they depict. Groups of strokes that are part of a stroke edge are no longer perceived as a single individual entity by becoming part of a stroke group. The individual strokes within such a group are separated by gaps, that is by groups of pixels whose grey-level intensity corresponds to the intensity of the image background. However, since these pixels are also part of the edge group, they are typically associated with the foreground rather than the background of the image. Thus, there are two different types of gaps present in a sketch, namely the *intra-group* gaps which are the gaps between edge strokes of the same stroke group, and the *inter-group gaps* which are the gaps between different stroke groups. In general, the inter-group gaps are larger than the intra-group gaps and this would allow us to distinguish between strokes that form part of an edge group and others that form part of different edge groups.

Bartolo et al. [2] note that the level of detail and the degree of the roughness of the sketched strokes may vary within the image and propose a multi-resolution approach through the use of Gabor filters (Box 2.2). Bartolo et al. use two filter banks: a quadrature filter bank with a bandwidth of one octave tuned to respond to the over-sketched line strokes and a center-off filter bank with a bandwidth of two octaves tuned to respond to the inter-group gaps. The maximal response at each pixel (x, y) is obtained from both filter banks and the result of the center-off filter bank is subtracted from the result of the quadrature filter bank. Thus, the center-off filter bank is used as an inhibiting filter to inhibit the quadrature response at the intra-group gaps as illustrated in Figure 2.2.

Donati et al. [13] use a similar multi-resolution approach but rather than using the Gabor filter framework, the image is processed using template matching via the Pearson correlation coefficient. The multi-resolution analysis is obtained by using 2D Gaussian kernels, each with a different variance value, as templates. Donati et al. then define the simplified sketch as:

$$S(x, y) = \begin{cases} R_{max}(x, y) & \text{if } |R_{max}(x, y)| > |R_{min}(x, y)| \\ R_{min}(x, y) & \text{otherwise} \end{cases} \qquad (2.1)$$

Box 2.2   Gabor Filters

The human primary visual cortex has, among others, simple cells which are sensitive to different spatial frequencies such that their response forms a tuning curve that peaks at some optimal frequency. This allows us to perceive a range of different spatial frequencies from around two cycles per degree to higher spatial frequencies of around 10 cycles per degree. These cells are also sensitive to an orientation such that their response peaks when patterns have the same combined frequency and orientation to which the cells are sensitive. Besides simple cells, the visual cortex also has complex cells which are phase invariant and hence provide a more continuous response. The action of these visual cortex cells can be simulated by Gabor filters [24].

The Gabor filter is a band-pass filter characterized by a sinusoidal signal modulated by a Gaussian function and is defined by Equation (a). and can, therefore, be tuned to respond to grating patterns of different frequencies and orientations as shown in Figure (i).

$$g_{\lambda,\theta,\psi,\sigma,\gamma}(x,y) = \exp\left(-\frac{x'^{\,2} + \gamma^2 y'^{\,2}}{2\sigma^2}\right)\cos\left(2\pi\frac{x'}{\lambda} + \psi\right) \qquad \text{(a)}$$

$$x' = x\cos(\theta) + y\sin(\theta)$$

$$y' = x\sin(\theta) + y\cos(\theta)$$

The filter is parameterized by the phase ψ, the wavelength λ of the sinusoidal signal, the aspect ratio γ the orientation θ and the standard deviation σ of the Gaussian function. Convolving an image with a Gabor filter of specific orientation and frequency allows for the selection of image regions that have grating patterns of the same characteristics as shown in Figure (i).

The Gabor filter can be made phase invariant by implementing it as a quadrature filter using the Gabor energy $G(x,y)$ defined by Equation (b).

$$G_{\theta,\lambda,\gamma}(x,y) = \sqrt{\hat{g}_{\theta,\lambda,\gamma,\phi=0}^2(x,y) + \hat{g}_{\theta,\lambda,\gamma,\phi=\pi/2}^2(x,y)} \qquad \text{(b)}$$

where $\hat{g}^2_{\theta,\lambda,\gamma,\phi}(x,y)$ is the response of the image to a Gabor filter with the parameter tuple $(\theta, \lambda, \gamma, \phi)$

The Gabor filter bandwidth is given by Equation (c) and the dependency of the filter's standard deviation σ on the wavelength λ ensures that the Gaussian envelope always contains the same number of sinusoidal cycles irrespective of the frequency selected.

$$b = \log_2 \left(\frac{\frac{\sigma}{\lambda} + \frac{1}{\pi}\sqrt{\frac{\ln 2}{2}}}{\frac{\sigma}{\lambda} - \frac{1}{\pi}\sqrt{\frac{\ln 2}{2}}} \right) \qquad \text{(c)}$$

The Gabor filter is used here to group over strokes for sketch simplification. In Section 5.4 the Gabor filters are re-introduced as feature descriptors for sketch-based shape retrieval.

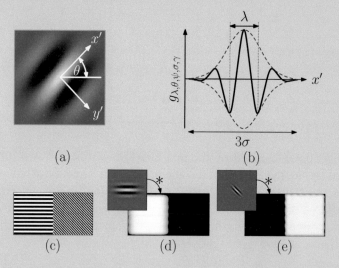

(a) (b)

(c) (d) (e)

Figure (i) The Gabor filter is characterized by sinusoidal signal modulated by a Gaussian function and can be tuned to match the orientation and frequency of orientation of grating patterns. Applying filters of different wavelengths and orientations will allow for the selection of different patterns in an image.

Quadrature Filter Bank

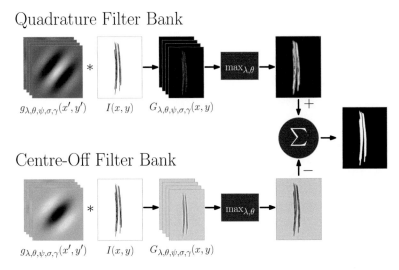

Centre-Off Filter Bank

Figure 2.2 Sketch simplification using Gabor filter banks. The simplification uses two filter banks, a quadrature filter bank which tuned to the over-sketched line strokes and a centre-off filter bank tuned to the inter-group gaps. The response of the center-off filter bank is used to inhibit the quadrature filter response.

$$R_{max}(x, y) = \max_{\forall i \in [0, N-1]} r(I, K_i) \tag{2.2}$$

$$R_{min}(x, y) = \min_{\forall i \in [0, N-1]} r(I, K_i) \tag{2.3}$$

where $r(I, K_i)$ is the Pearson correlation coefficient between the image I and the Gaussian kernel K_i and N is the number of Gaussian kernels. Equation 2.1 effectively groups line strokes and intra-group gaps while inhibiting inter-group gaps.

Rather than using hand-crafted kernels, Simo-Serra et al. [43] achieve multi-resolution processing of the image through the use of convolutional neural networks (CNNs). The network is divided into three parts. The first part acts as an encoder which spatially compresses the image. In so doing, fine over-sketched strokes are sub sampled and aggregated together. The second part extracts the essential lines from the image while the third part acts as the decoder that brings back the simpler representation of the drawing back to its full-sized gray-scale image. The network architecture used is similar to the U-Net architecture, but rather than using max-pooling, the architecture uses convolutions with a stride > 1 for down-sampling the image and a stride < 1 for up-sampling the image. The model is trained using pairs of

---

Box 2.3   Template matching with the Pearson Correlation Coefficient

Given an $R \times C$ image I and a $(2h + 1) \times (2w + 1)$ template T, where $(2h + 1) < R$ and $(2w + 1) < C$, the normalised Pearson Coefficient at a point (x, y) in the image is given by:

$$r(I, T, x, y) = \frac{\displaystyle\sum_{i=-w}^{w} \sum_{j=-h}^{h} (I(x+i, y+j) - \bar{I}_{ij})(T(i,j) - \bar{T})}{\sqrt{\left(\displaystyle\sum_{i=-w}^{w}\sum_{i=-h}^{h}(I(x+i,y+j)-\bar{I}_{ij})^2\right)\left(\displaystyle\sum_{i=-w}^{w}\sum_{i=-h}^{h}(T(i,j)-\bar{T})^2\right)}}$$

where I_{ij} is the windowed image region of size $(2h + 1) \times (2w + 1)$ and centered at (x, y) while $\bar{I}_{ij}$ and $\bar{T}$ are the means of the windowed image and the template respectively. The denominator of r normalizes the Pearson correlation coefficient to the range $[-1, 1]$ where a coefficient of -1 indicates a total negative correlation and a coefficient of 1 indicates a total positive correlation.

---

rough sketches and their simplified counterparts using the weighted mean square error criterion, given by Equation 2.4 as the loss function,

$$L(y, \hat{y}, M) = ||M \odot (y - \hat{y})||^2 \tag{2.4}$$

where $\hat{y}$ is the simplified image generated by the simplification network, given the input image x, y is the ground-truth image, M is a loss map and $\odot$ is the matrix element-wise multiplication operator. Simo-Serra et al. set the loss map to reduce the loss on thicker lines such that the model does not focus on these thicker lines at the expense of thinner lines. The local, normalized histogram at each image pixel is used as an indicator of the thickness of the line strokes in the local region, defining the loss map as:

$$M(u, v) = \begin{cases} 1 & \text{if } I(u, v) = 1 \\ \min(\gamma \exp(H(I, u, v) + \kappa, 1) & \text{otherwise} \end{cases} \tag{2.5}$$

where $H(I, u, v)$ is the bin value of the histogram in which the pixel $I(u, v)$ falls into while γ and κ are weight parameters. The loss function, therefore, gives a higher loss value to pixel locations falling within darker local regions.

To allow the CNN to generalize to natural sketches and also to make use of the availability of uncoupled rough and neat sketches, Simo-Serra et al. [44] extend this sketch simplification network to one which involves an adversarial augmentative framework based on the principles of generative adversarial networks (GANs) (see Box 2.4). However, Simo-Serra et al. note that for sketch simplification the output $\hat{y}$ generated by the network must be predicted from the sketched input x and that for the generated image to have resemblance to the input image, a loss function that measures the similarity between the generated image and the intended ground-truth is needed. Thus, Simo-Serra et al. propose to use a prediction model $S : x \mapsto \hat{y}$ which is trained jointly with the discriminator network $D : y \mapsto D(y) \in \mathbb{R}$ where the training is conditioned by the loss function $L(y, S(x))$ where y is the ground-truth image corresponding to the input image x, thereby creating *supervised adversarial training* defined by the expectation value over a supervised training set $\rho_{x,y}$ of input-output pairs as optimizing the expression:

$$\min_{S}\{\max_{D}\{\mathbb{E}_{(x,y)\sim\rho_{x,y}}[\alpha \log(D(y))+$$
$$\alpha \log(1 - D(S(x))) + L(S(x), y)]\}\} \quad (2.6)$$

The term α is a weighting hyper-parameter that controls the influence of the adversarial training. Set too low, the influence of the adversarial training will become negligible. Set too high, the influence becomes such that the generated images, while realistic, do not resemble the input. Simo-Serra et al. further note that the first two terms of the expectation values are not dependent on the joint input-output image pairs and, given that decoupled rough and neat drawings can be sourced more easily than corresponding pairs, propose a further modification to the optimizing function changing it to:

$$\min_{S}\{\max_{D}\{\mathbb{E}_{(x,y)\sim\rho_{x,y}}[\alpha \log(D(y))+\alpha \log(1-D(S(x)))+L(S(x),y)]+$$
$$\beta\mathbb{E}_{y\sim\rho_y}[\log(D(\hat{y}))] + \beta\mathbb{E}_{x\sim\rho_x}[log(1 - D(S(x)))]\}\} \quad (2.7)$$

The term β is a second weighting hyper-parameter. Note that the model loss term $L(S(x), y)$ remains critical in training ensuring that the generated images remain anchored to the input images while the unsupervised data provides the flexibility of exposure to a larger variety of sketched drawing styles.

The simplification network used here remains the same as that described in [43]. However, the loss function is simplified to:

$$L(S(x), y) = ||S(x) - y||^2 \quad (2.8)$$

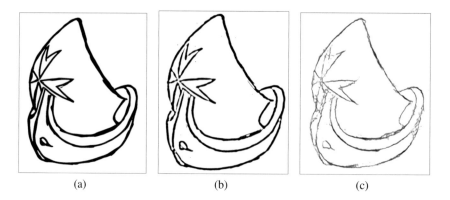

(a)   (b)   (c)

Figure 2.3   The simplification of the sketch shown in Figure 2.2(b) using (a) Gabor filter banks proposed in [2], (b) template matching proposed in [13] and (c) Convolutional neural networks proposed in [43]

Thus, the loss function is no longer weighted by the thickness of the strokes. The discrimination network used is a "shallow" network consisting of seven layers, the last of which is a fully-connected layer. The first layer uses 5×5 convolutions with 16 filters and all subsequent layers use 3×3 convolutions and double the number of filters per layer. All fully connected layers, bar the last layer, use a rectified linear unit (ReLU) activation function. The last layer uses the sigmoid activation function such that the network has a single output corresponding to the probability of the input belonging to the real data ρ_y rather than artificially generated $S(x)$.

Figure 2.3 gives a comparative example of the three simplification methods described above, using $\lambda = [18/1, 18/2, 18/4, 18/8]$ and $\sigma = 0.56\lambda$ for the Gaussian parameters in both the Gabor-filter approach and the template-matching approach. We note that in this particular example, the CNN-based approach does not simplify all over-strokes equally and that is particularly true when over-sketched line strokes merge at junctions. The Gabor-based and template-matching simplification both provide better results. In this particular example, the template-matching approach provides thinner line strokes and, in instances, better suppression of inter-group gaps than the Gabor-simplification. However, this is at the expense of the suppression of some line strokes.

These techniques create simplified raster images from the rough sketches and require further post-processing to obtain the vector strokes. However, the simplification of the drawing from the over-sketched lines to single lines

---

Box 2.4   Generative Adversarial Networks

In a generative adversarial network framework, two network models, namely a generative model and a discriminative model are jointly trained such that, given a random input, the network can stochastically generate an output drawn from the distribution of the training set ρ_y. The role of the generative model can be described as the neural network $G : z \mapsto y$ that maps the random input z to the output image y. The discriminative network $D : y \mapsto D(y) \in \mathbb{R}$ on the other hand computes the probability that an image y is obtained from real data rather than generated by the generative network. The two networks are jointly optimized with respect to the expectation value:

$$\min_{G}\{\max_{D}\{\mathbb{E}_{y\sim\rho_y}[\log(D(y))] + \mathbb{E}_{z\sim\rho_z}[\log(1 - D(G(z)))]\}\} \quad \text{(a)}$$

By alternatively maximizing the classification log-likelihood of D so that it can discriminate between real and generated images, and then, optimizing G to deceive D makes it possible to train the generative model to create realistic images from random input.

GAN has been extended to conditional generative adversarial network (CGAN) where the generative model takes an additional input such that $G : (x, z) \mapsto y$ generates the output image y from the random input z and the additional input x. Likewise, the discriminative model $D : (x, y) \mapsto D(x, y) \in \mathbb{R}$ also takes the input x as an additional input to evaluate the conditional probability that the image y is real or generated, given the input x.

Both GAN and CGAN can, therefore, be used to generate realistic synthetic data that can be used for training purposes.

---

would make the vectorization process more straightforward. The techniques used adopt convolution-based processing, judging the group or otherwise of line strokes based on the local neighborhood. The size of this neighborhood is determined from the bandwidth of the filter kernels in Bartolo et al. [2] and Donati et al. [13] and the number of convolutional layers in Simo-Serra et al. [43, 44]. Although the latter two approaches take into consideration a larger neighborhood than the former two approaches, the processing can still be considered as local-based. Further simplification can be achieved by taking into consideration the global properties of the sketch, such as the perceptual ability and preference to form smooth contours or closed loops as described by Gestalt principles. While the algorithms discussed here do alter the image to the extent of grouping like strokes together, they do not have the ability to bridge over gaps brought about by drawing incompleteness. Algorithms such as those proposed by Liu et al. [27] and Liu et al. [28] do take Gestalt principles into consideration, but these algorithms assume that the stroke vector data is already known, that is, they are suitable for online sketching applications not to raster-born images.

In Sasaki et al. [40], a CNN approach is used to bridge gaps along the contour of a sketch. The goal of the network is then to detect the missing parts on the contour and fill these missing parts realistically and coherently to close the strokes. The filled in section is intended to match the stroke characteristics including conserving both thickness and curvature. The model architecture used follows the encoder-decoder framework, with the first half extracting features from the image by reducing the image resolution, while the second half consolidating the lines and sharpening the output. A ReLU is used as the activation function for all by the last convolutional layers. For the last layer, a sigmoid function is used, ensuring that the output gray-scale image is normalized within the range $[0, 1]$. The network is trained by using images with different line styles for which random patches of varying sizes are erased. The mean square error is used as the model loss for learning. While the approach does perform well to close relatively short gaps along the line contours, the algorithm does not perform as well when the gap in the contour occurs at the junction region between multiple lines. This performance can be explained by the complexity and multiple combinations possible at junction regions, which, as discussed above, requires a global understanding of the structure of the drawing.

2.1.2 Improving Junction Point Representation

Improvements in the localization and representation of the line strokes at junctions have been tackled through techniques that use the topological structure of the drawing, hence taking into consideration the global structure of the drawing. Noris et al. [33], Bessmeltsev and Solomon [3] and Favreau et al. [17] all use a graph-structure approach to represent the topology of the drawing. These methods take as input, a relatively neat drawing, possibly

> **Box 2.5   Minimum spanning tree**
>
> A minimum spanning tree is a connected edge-weighted, undirected graph that connects all graph vertices together without any cycles and with the minimum possible total edge weight.

with some over-sketching, but where the over-sketching is fairly limited to the intended stroke region with a very clear distinction between inter-group and intra-group gaps.

In Noris et al. [33], rather than finding the image skeleton from the binarised image, the intensity-gradient is used to move pixels in the direction of the gradient until they aggregate at the center-line. In so doing, an estimation of the stroke width is obtained. A minimum spanning tree (see Box 2.5) is created from the clustered pixels, connecting each aggregated pixel to the other and using the distance between pairs of pixels as the weight of the spanning tree edges. Branches of the spanning tree with lengths smaller than the line width are removed such that the resulting tree approximates the image skeleton. Since the minimum spanning tree does not allow any cycles, any closed loop in the drawing will be broken. Thus Noris et al. propose to perform a local minimum spanning tree at each leaf node as well as a global minimum spanning tree, merging the two to restore the loop. The path that forms the shortest junction-point to end-point distance is considered as the baseline skeleton of the drawing. However, at junction regions, the base-line skeleton may not be completely accurate and Noris et al. propose a reverse-drawing step. At each potential junction location, a circular mask is grown until the strokes no longer overlap. This finds the ambiguous region in the drawing. The baseline centers within this region are removed and replaced with all possible configurations of continuous center-line candidates that connect the line strokes are created as illustrated in Figure 2.4. The straightest of the connections is retained, thereby emulating the Gestalt law for smooth continuation.

Favreau et al. [17] use a region-based approach, namely the trapped-ball segmentation algorithm, to determine the skeleton of the sketch. Through this approach, the skeleton is the boundary between two adjacent regions. The topology graph of the sketch is created from this skeleton with the nodes v being the junctions or endpoints and the edges E the skeleton branches between the junctions. Each edge $e \in E$ is represented as a Bézier curve segment B^e. Favreau et al. acknowledge that the skeleton will

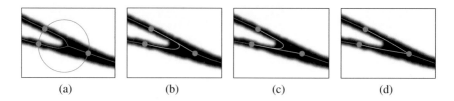

(a)	(b)	(c)	(d)

Figure 2.4   Adjusting the vectors at a suspected junction position. The baseline centres inside a suspected junction region are removed (a) and possible configurations of continuous centre-line candidates that connect the line strokes are created (b-c). The smoothest path is then selected as the vector representation at the junction.

cause over-segmentation as a result of the poor definition of the skeleton towards the junction regions. Favreau et al. propose that to resolve such over-segmentation, a balance between fidelity of the parametric curve to the input drawing and the simplicity of the resulting vector graphics should be reached. Favreau et al. propose the use of hyper-graphs to reduce the complexity of the vectorization, removing the extra branches at junctions by joining adjacent edges into hyper-edges. The goal is, therefore, to explore the possible set of hyper-graphs finding that which offers the best trade-off between simplicity of the network and fidelity to the input drawing. Fidelity is measured by computing the error between the Bézier fit and the sketched stroke. Simplicity is determined by two factors, namely the order of the Bézier curve and the number of curve segments that represent the skeletal segment. Favreau et al. explore the hyper-graph space using the Metropolis-Hastings search algorithm which makes random explorations by iteratively perturbing the graph structure with the perturbed structure becoming the actual structure conditions on the energy variation between the two. Favreau et al. allow three perturbations: hyper-edge merging and splitting, switching the order of the Bézier curve and the overlap between hyper-edges.

Like Noris et al. [33], Bessmeltsev and Solomon [3] start with a gray-scale image but, rather than aggregating strokes towards the baseline center to form the node vertices, use stroke bundles to form the topology graph. The stroke bundles are determined with the help of the frame-field computed at or near the pixels forming the line strokes. This frame-field consists of two vectors (u, v) which are aligned to the tangent direction of the curve. At junction regions, the frame-field aligns to the directions of both curves are involved in the junction. Bessmeltsev and Solomon constrain the estimation of the frame-field such that it is aligned with the gradient direction estimated from the Sobel gradient g. They also constrain the frame-field to be smoothly continuous. In so doing, the frame-field may be determined through an optimization problem. Curve bundles are formed by grouping curves with the same frame-field direction, with the bundle being represented as a vertex on the edge-vertex graph. Starting with an initial seed-point on the curve, the frame-field direction is used to propagate the seed along the curve, thereby adding edge-vertex pairs to the topology graph. Since the curve boundary may be uneven, the topology graph may contain additional loops or branches. Any such loops or branches shorter than some pre-established lengths are pruned, thus simplifying the topology. As lines approach the junction, they merge in parallel or near parallel directions. Such regions can be detected from valence-2 graph nodes connecting valence-3 nodes. In such cases, the graph is split at the valence-3 nodes, replicating the strokes between these nodes, thus decoupling the grouped strokes at the junction region into separate line strokes. The final vectorization is formed by taking the centers of the stroke bundles, creating an auxiliary graph, adjusting the nodes at junction points such that the resulting position balances curve smoothness and remaining close to the stroke centers.

The results obtained by the three techniques are similar when neat drawings with no shading, gaps, and very little over-sketching is present. Bessmeltsev and Solomon [3] produce the tightest junctions of the three and this is particularly important to capture folds and creases at more organic object shapes. Bessmeltsev and Solomon [3] are also better than Noris et al. [33] at removing smooth shading and producing single lines in the presence of over-traced strokes, although both algorithms capture shading when this is present at hatching marks. Favreau et al. [17] are also able to produce single lines from over-tracing and has similar outcome in the presence of shading. The algorithm, however, tends to over-simplify the drawing and can lose significant detail, particularly in the presence of gaps in the drawing. This is a direct result of the skeletonization approach adopted. If no gaps are present in the drawing reasonable simplification can be achieved in the presence of both shading and over-sketching.

In the above, junction localization was described within the context of sketch interpretation. However, it is worth noting that similar junction localization is used in other computer vision applications, such as within the field of biometrics, for example, in the detection of minutiae in fingerprints or bifurcations in retinal vascular images among others. Here, localization of the minutiae or bifurcations, without the need to model the ridges or vascular segments, is often sufficient and can be achieved through template matching, using, for example, the hit-and-miss transform [9, 4]. Such approaches typically assume that with good pre-processing, it would be possible to obtain fingerprint ridges or the retinal vascular network of sufficient quality to allow for simpler approaches similar to those adopted in earlier vectorization approaches of Tombre et al. [50] and others described in Section 2.1.1. It is, however, possible to have degraded images where, for example, fingerprint ridges are observed [51]. In such cases, besides the detection of the existing minutiae, it is also necessary to restore the image. Tu et al. [51] proposed to do so using cubic Bézier curves to model existing ridgelines and introduce domain knowledge to deduce the connectivity between these lines where the image is obscured. This approach is indeed similar to that introduced by Favreau et al. [17] using the natural formation of the ridgelines as the driving domain knowledge rather than the Gestalt laws of perception. Thus, although the application areas may differ, it is possible to cross over ideas from one domain to another. Indeed Bonnici et al. [6] adopt the combination of shifted responses (COSFIRE) used to detect bifurcations in retinal images with the topology of planar objects to detect junction positions in relatively neat drawings with shadows.

2.2 Curve Fitting

After locating all sketch strokes in the drawing, the final step of the vectorization process involves the representation of these strokes with meaningful segments. A simple approach would be to treat each vector point from the detected strokes as a critical point of the underlying sketch, fitting a curve that passes through each vector point. This approach, however, has two drawbacks. First, it retains too much unnecessary information which will have an impact on subsequent processing and rendering of the sketched strokes. The second drawback relates to the sensitivity of the vectorization to perturbations on the hand-drawn strokes. Keeping all points as critical points retains all these perturbations, whereas smoother curves would not only be more visually pleasing but most likely closer to the user intent. The final step of the vectorization process is, therefore, a curve-fitting problem that aims to find smooth curves or line segments to represent the located vector points [50]. Thus, the curve-fitting problem requires the resolution of two conflicting goals. On the one hand, we want to find compact descriptors such that any high-level algorithm can work with the resulting descriptors efficiently. On the other hand, we want to represent the sketch as accurately as possible, that is, we want to minimize the error between a curve constructed from its descriptors and the original drawing [35].

A polygonal approximation is perhaps the simplest approach to curve-fitting. This approach represents the curve as piece-wise linear segments between critical points. The question that arises is, therefore, how to determine the locations of the critical points on the digital curve. For a curve expressed as $y = f(x)$, such critical points would typically be those points with high curvature, where the curvature k at a point (x, y) on the graph is defined as

$$k = \frac{y^n}{\sqrt[3]{1 + y^{2^2}}} \qquad (2.9)$$

and the critical points would then be those points with maximum curvature [42]. Equation 2.9 assumes continuous curves and so, for discrete, digital curves, approximation methods are required to evaluate the location of critical points.

The optimal segment finder described by Sklansky and Gonzalez [46] finds the critical points $c = \{v_1, \cdots, v_k, \cdots, v_N\}$, from the list of vector points $V = \{v_1, \cdots, v_N\}$ such that $|c| < N$, thus forming an idealized polygonal curve S by joining pairs of consecutive critical points. The critical points are such that S lies within a distance ϵ from all vector points. Sklansky and Gonzalez propose the Hausdorff-Euclidean distance $H(V, S)$ as the distance metric.

A naive approach to locate the critical points would be to start with the last identified critical point v_k and set the subsequent vector point v_{k+i} as the next critical point, starting with $i = 1$ and incrementally increasing the value of i until the line segment $\overline{v_k, v_{k+i}}$ remains within $|\epsilon|$ of the vector points $\{v_p | p = k, \cdots, k+i\}$. This approach, however, requires the re-computation of the Hausdorff distance for each increment of i, which becomes inefficient for long line segments. Instead, Sklansky and Gonzalez suggest a more efficient implementation that places a circle of radius ϵ onto each vector point. Starting from a critical point v_k, they then construct a cone to the next candidate critical point v_{k+i}, by tracing rays from v_k tangent to the circle at v_{k+i} as illustrated in Figure 2.5. A candidate point v_{k+i} becomes a critical point if one of the intermediary points between v_k and v_{k+i} lies outside this cone.

The parameter ϵ determines the coarseness of the polygonal fit. Larger values of ϵ approximate the sketched curve with longer line segments at the expense of missing or displacing critical junction or turning points. On the other hand, smaller values of ϵ improves the detection of the critical points at

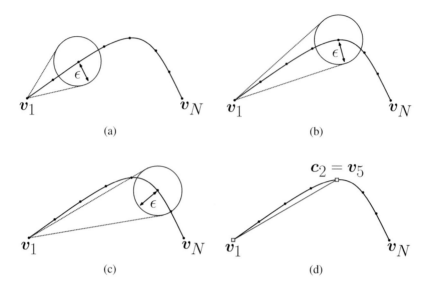

(a)

(b)

(c)

(d)

Figure 2.5 (a) A circle with radius ϵ is drawn around the vector point and a cone is formed by drawing tangents to the circle from the first critical point. (b) The cone narrows as more points are added to the list of vector points until (c) a vector point is added such that one or more vector points lie outside the cone of support. (d) The previous vector point is then selected as the next critical point and the curve is approximated by a straight line joining the two critical points.

the expense of capturing too many fine details on the sketched curve, possibly over-segmenting the curve. Debled-Rennesson et al. [10] resolve the problem associated with the selection of a single threshold parameter by proposing a multi-order analysis framework. Debled-Rennesson et al. note that the vector points represent an underlying sketched line with width w and define this line as $D(a, b, \mu, w)$ where a/b is the slope of the line, μ is the line's lower bound and w its thickness. The vector point $v_k = [x_k, y_k]$ represents a point on this line and satisfies the relation:

$$\mu \leq ax - by < \mu + w \tag{2.10}$$

as illustrated in Figure 2.6. Debled-Rennesson et al. further define the order d of the line such that $d \geq \frac{w}{\max(|a|, |b|)}$.

Starting with an initial critical point v_k and given a line order d, consecutive vector points v_{k+i} can be added to the line as long as they satisfy Equation 2.10. Similar to the use of the ϵ parameter in Sklansky and

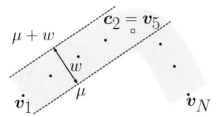

Figure 2.6 Illustrating the blurred-segment curve approximation [10]: A line segment of order d can be defined by the lower bound μ and and upper-bound $\mu + w$. Starting with an initial line fragment, subsequent vector points are added to the line segment as long as they fall within the line's upper and lower bounds. The last vector point that falls within the line segment becomes the next critical point.

Gonzalez [46], the line order d provides a trade-off between critical point localization and over-fitting of the sketched lines. Debled-Rennesson et al., however, make a direct link between d and the line thickness, endowing meaning to the parameter since thicker lines do have more tolerance on the placement of the critical points. Moreover, since d is typically an unknown parameter, Debled-Rennesson et al. compute the line segment approximation using different values of d, retaining the largest value of d for which the number of line segments does not increase further. The location of the critical points is then refined by back-tracking to the lowest order of d.

These two curve-approximation methods add data points from the existing digital curve to its approximation sequentially, until no other data point remains. Such approaches are fast and straightforward to implement but can be dependent on the chosen starting point on the curve, particularly if the curve forms a closed contour. A different approach to the polygonal approximation has been adopted by Rosin and West [39] who in turn, adapt an earlier approach introduced in Lowe [29]. Here, the authors consider the curve in its entirety, splitting the curve recursively at the maximum deviation point from the approximating straight line. Each split-level requires a decision to determine whether a single straight line is better than a lower descriptor which describes the curve with two or more line segments. Rosin and West propose to use a goodness-of-fit or significance value, defined as the ratio $error/length$, to determine the suitability of the line representation. Rosin and West note two advantages of using this ratio. The first is that this eliminates the need for a pre-set threshold value. More importantly, the significance ratio allows more error over longer curves, ensuring that curves of different lengths get similar approximations but at different scales as illustrated in Figure 2.7.

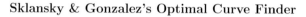

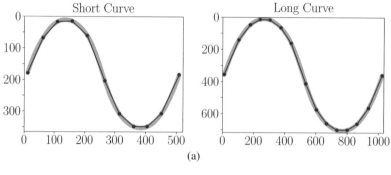

(a)

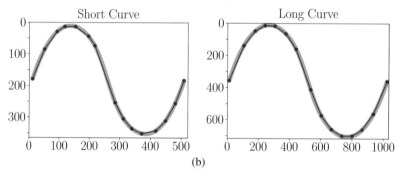

(b)

Figure 2.7   Showing the difference between the Optimal Curve Finder approach [46] and the use of the $error/length$ approach [39]. The curves in on the left are shorter representations of the curves on the right. While this example shows that a similar curve approximation is achieved by both approaches for the longer curves, it shows that the $error/length$ ratio provides for a more accurate representation of the shorter curve.

It is also possible to determine the critical points using metrics other than distance. For example, Shao and Zhou [42] utilise the area between fixed-length chords and curve formed by the full vector point set, as illustrated in Figure 2.8. At critical points, the underlying curve segment deviates from the fixed-length chord, resulting in a larger area between the two. The area metric, therefore, indicates the neighbourhood around the critical point. If the chords are sufficiently small, it is possible to narrow down the neighbourhood to a single point by taking the center of the chord. An alternative, more accurate approach would be to fit the maximum inscribed triangle between the fixed-length chord and the curve defined by the vector-points. The apex

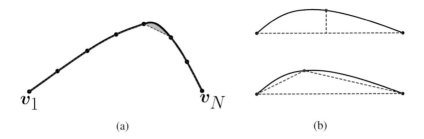

v_1 v_N

(a) (b)

Figure 2.8 (a) The area between fixed length chords and the curve gives an indication of the location of the critical points. (b) the position of the critical point can be estimated from the center of the chord, or, more accurately, from the apex of the maximally inscribed triangle

of this maximum inscribed triangle will coincide with the critical point of the curve.

Pal et al. [35], on the other hand, propose to use the chain-code as the metric to locate critical points. The chain-code numbers the eight principal directions of movement on the compass rose. With such a code, we may represent the sequence of vector points with a number which indicates the direction of movement while traversing the curve. Pal et al., then label a vector point as a critical point if the change in direction between two vector points $C(V_k)$ and $C(V_{k+1})$ is greater than one. The chain-code, however, is sensitive to any perturbations on the sketched curve. In this case, the problem appears to a lesser extent since Pal et al., apply the chain-code to the vector points sampled from the curve rather than directly on the sketched strokes. Moreover, to further mitigate the problem of over-fitting, Pal et al., adopt a split-and-merge approach similar to that introduced in [39]. Thus, Pal et al., propose to merge two curve segments by deleting the critical point if $D(v_k, v_{k+1}) + D(v_{k+1}, v_{k+2}) > D(v_k, v_{k+2})$ and $D(v_k, v_{k+2}) < \epsilon$ where $D(a, b)$ is the error function and ϵ is a threshold on the goodness of fit.

From the discussion thus far, we may note that sequential approaches tend to place critical points at the first, non-conforming vector point. In contrast, iterative methods tend to consider the curve in its entirety to determine the best placement of the critical point. Optimization algorithms can also find the optimal placement of the critical points, albeit at an increased computational effort.

Di Ruberto and Morgera [11] adopt one such approach using ant-colony optimisation to optimise the placement of critical points on closed-loop curves. The approximating curve is expressed as the graph $G = (V, E)$ with nodes V representing the critical points and the edges E the line segments connecting the critical points. Ants traverse the graph by moving from one critical point to another to form a path, where the length of the path is equivalent to the number of critical points on the path. Di Ruberto and Morgera define the path fitness in terms of the length of the path and the error between the approximating path and the vector points. The goal of the optimization is to find the shortest path, which also minimizes the error between the path and underlying vector points.

Initial placement of the ants can follow a random distribution. However, placing the ants closer to the expected critical points improves the chances of finding the ideal path. Di Ruberto and Morgera propose two variations on the initial ant placement. The first uses the shape signature as the shape descriptor to locate the extrema points, selecting the corresponding nodes on the graph as starting points. The second selection strategy starts by evaluating the largest approximation error among all directed edges starting at a node i. Sorting the nodes in order of increasing error allows the selection of the first D nodes as the starting nodes.

In both cases, in subsequent cycles, ants favour those nodes with a higher probability of forming short closed contours, expressing this probability as the trade-off between the average quality of solutions chosen by ants at starting node i and the prior probability of the node. Moreover, Di Ruberto and Morgera introduce a heuristic measure to control the transition between nodes. This heuristic encourages ants to walk to the furthest node possible in order to create polygons with long segments and fewer vertices.

Using a similar optimization philosophy, Alvarado-Velazco and Ayala-Ramirez [1] cast the problem of finding critical points in a genetic algorithm. By modeling the chromosome as the index of vector points, the scope of the genetic algorithm becomes that of finding the node indices which are best candidate critical points. Alvarado and Ramirez base the fitness function which controls the evolution of the population on heuristics which define features for good quality curve approximations. These include the distance from the vector-points, collinearity, vertex over-crowding, the bounding box factor and length similarity. Alvarado-Velazco and Ayala-Ramirez observe that the length similarity and bounding box factor ensure that the position of the critical points evolves to form a curve which is similar in length and shape as the underlying vector-points. Moreover, the vertex over-crowding feature ensures that critical points are evenly spread along the curve, thus ensuring consistency in detail across the curve.

In the discussion thus far, it was assumed that critical points would be joined by straight line segments to form piece-wise linear curves. While such a representation is suitable for drawings of objects with linear boundaries, this representation causes over-fragmentation in sketches consisting of a mixture of linear and curved boundaries. In such cases, higher-order curve-fitting on the vector points results in a better curve-fit.

In Rosin and West [39] sequences of line segments from the polygonal approximation are hypothesised as circular arcs, thus requiring evaluation of four arc parameters, namely the centre coordinates, the start and endpoints of the arc and the arc's radius. Fixing the start and endpoints of the circular arc to the endpoints of the polygonal sequence reduces the unknown parameters to two. Moreover, Rosin and West propose to rotate and shift the arc segment to a temporary axis such that the chord joining the two outer points of the arc lies on the x-axis, centred on the origin. In this way, the centre of the circular arc lies along the y-axis further reducing the unknown quantities. The arc-fit error can then be expressed as the error $\sum_{i=2}^{N-1} e_i^2$ where N is the total number of vertices of the polygonal approximation, and e_i is the shortest distance to the circular arc. The radius, and hence, the position of the centre can then be determined through a recursive binary split algorithm, using gradient descent to search for the minimum error. Rosin and West, use a similar recursive process used for straight lines, splitting of the circular arc into subarcs until the number of critical points is not less than four, the limit for over-fitting circular arcs. At the end of each iteration, the circular arc replaces the straight-line representation only if it is a better fit for the digital curve.

While circular arcs can represent curved lines better than straight line segments, not all curves are well-represented by circular arcs. Indeed, some sketched lines can be better represented by more generic curves. Linear regression may be used to fit higher order polynomials to the sketched strokes [45]. Care, however, must be taken to ensure the continuity between a curve segment and the next in order to maintain smooth continuity along the contour. It is then desirable to have additional control over the start and end fragments of the fitted curve. One such family of curves are Bézier curves (Box 2.6). Since Bézier curves are defined by their control points, P, we can describe the curve-fitting problem as that of finding the control points P from which we can create a Bézier curve such that the points p on the curve lie close to the vector points v obtained from the drawing. That is, we want to find control points which minimise the error

$$E(\boldsymbol{P}) = \sum_{i=1}^{n} (\boldsymbol{p}(t) - \boldsymbol{v}_i)^2 \tag{2.11}$$

Since the Bézier curve is defined by control points, we can describe the curve-fitting problem as finding the control points $\boldsymbol{P}$ from which we can generate curves that are as close as possible to the vector points $\boldsymbol{v}$ obtained from the drawing.

The solution to Equation 2.11, however, cannot be solved using least-squares techniques [42]. Instead, Bézier-fitting techniques use initial approximations to find the optimal control points. Schneider [41] and Shao and Zhou [42] use the tangent property of the Bézier curve to find the position of the control points. They note that given the fixed positions of the first and last control points, the intermediary points of a cubic Bézier can be expressed as, $\boldsymbol{P}_1 = \boldsymbol{P}_0 + \alpha_1 \boldsymbol{t}_0$ and $\boldsymbol{P}_2 = \boldsymbol{P}_3 + \alpha_2 \boldsymbol{t}_1$, where $\boldsymbol{t}_0$ and $\boldsymbol{t}_1$ are the unit tangent vectors at $\boldsymbol{P}_0$ and $\boldsymbol{P}_1$ respectively while α_1 and α_2 are scalar values that indicate the distance of the intermediary control points from the curve end-points.

Schneider [41] approximates the tangents by fitting a least-squares line to the points at the start and end of the vector-point list, thus, obtaining $\hat{\boldsymbol{t}}_0$ and $\hat{\boldsymbol{t}}_1$. With this approximation of the tangent-pair, the minimization problem reduces to solving

$$\frac{\partial E(\boldsymbol{P})}{\partial \alpha} = 0 \tag{2.12}$$

Schneider demonstrates that this can be solved linearly to obtain α_1 and α_2, thus finding the intermediary control points. Before doing so, however, it is necessary to register the vector points with their position on the curve. Schneider uses chord-length parameterization to achieve this, such that the values of α_1 and α_2 are estimates, dependent on the roughness of the chord-lengths. Schneider then applies Newton-Raphson iterations to improve the location of the control points.

Shao and Zhou [42] adopt a similar approach, but rather than approximating the tangent pair through least-square fitting they first set initial values for $\hat{\boldsymbol{t}}_i$ and solve for α_i as described in [41]. They then use the α_i obtained to find an improved solution for $\hat{\boldsymbol{t}}_i$, repeating the process until the error between the curve and the vector points is a minimum.

Pal et al. [35], take a different approach to estimate the initial position of the control points and to their subsequent refinement. Using De Casteljau's

Box 2.6   Bézier Curves

The Bézier curve is a parametric curve introduced in the 1960s by Pierre Bézier in the design of Renault cars. Bézier curves are related to the Bernstein polynomials:

$$B_i^n(t) = \binom{n}{i} t^i (1-t)^{n-i} \tag{a}$$

where n is the order of the polynomial,

$$\binom{n}{i} = \frac{n!}{(n-i)!i!} \tag{b}$$

and $t = [0, 1]$. A point $\boldsymbol{p}(t)$ on the Bézier curve is given by:

$$\boldsymbol{p}(t) = \sum_{i=0}^{n} B_i^n(t) \boldsymbol{P}_i \tag{c}$$

where $\boldsymbol{P}_o, \cdots \boldsymbol{P}_n$ are control points of the curve. The first and last control points typically correspond to the endpoints of the curve while intermediary control points do not necessarily lie on the curve. Together, the control points form a polygon which encloses the curve as illustrated in Figure (i). Equation c, therefore, interpolates the curve values between the control points. We can also note that the line segment joining the first two control points $\overline{\boldsymbol{P}_o \boldsymbol{P}_1}$ and the last two control points $\overline{\boldsymbol{P}_{n-1} \boldsymbol{P}_n}$ are tangents to the curve at $\boldsymbol{P}_o$ and $\boldsymbol{P}_n$ respectively as illustrated in Figure (i) [52].

To draw a Bézier curve with control points $\boldsymbol{P}_i$ we may apply Equation c for all values of t. However, an alternative approach, would be to use *De Casteljau's algorithm*. Here the Bézier is split recursively into two sub-curves at the parameter point t until we can estimate the desired point as the interpolation on a single straight line. That is, we can compute the Bézier curve $B_i^n(t)$ as an interpolation from two, lower-order Bézier curves using:

$$B_i^n(t) = (1-t)B_i^{n-1}(t) + tB_{i-1}^{n-1}(t) \tag{d}$$

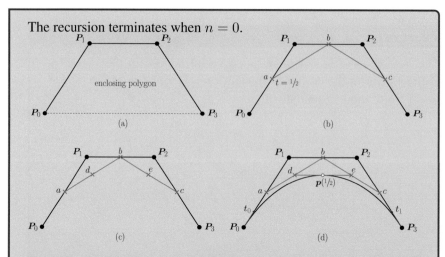

Figure (i)   De Casteljau's curve plotting. (a) The control points of the curve form an polygon which encloses the curve. Using the $t = 1/2$ point as an example: (b) The intermediary points a, b and c are formed by linear interpolation on the lines $\overline{P_0P_1}$, $\overline{P_1P_2}$ and $\overline{P_2P_3}$ respectively. (c) Two further points d and e are obtained by interpolating lines $\overline{ab}$ and $\overline{bc}$ (d) The point $p(1/2)$ on the curve is then obtained by interpolating the line $\overline{de}$. Note that the tangents curve at P_o and P_1 are defined by the enclosing polygon.

recursion, the point on the cubic Bézier curve at $t = 1/4$ and $t = 3/4$ are

$$p(1/4) = 1/64(27P_0 + 27P_1 + 9P_2 + P_3) \tag{2.13}$$
$$p(3/4) = 1/64(P_0 + 9P_1 + 27P_2 + 27P_3) \tag{2.14}$$

For the first approximation, Pal et al., assume that the curve intersects the vector points at the $t = 1/4$ and $t = 3/4$ points such that $p(1/4)$ and $p(3/4)$ are obtained directly from the vector point. Using these two points, Pal et al., obtain the first approximation of the intermediary control points. These coordinates are bit-encoded and represented as a chromosome. A genetic algorithm is then used to optimise the position of the control points, using $F(C) = R - E(C)$ as the fitness function, where

$$R = \frac{E_{max} - E_{min}}{(Selectivity - 1) + E_{max}} \tag{2.15}$$

$E(C)$ is the fitting error of the curve generated using the chromosome C and the E_{max} and E_{min} are the errors of the best fitting and worst fitting chromosomes of that generation's population and *Selectivity* is the ratio of the E_{max} and E_{min}. Pal et al., generate the first population by sampling a window around the initial approximations of the control points and picking random points from within these windows.

Masood and Ejaz [30], also use De Casteljau's recursion but note that the peaks of the cubic Bézier curve occur at $t = \frac{1}{3}$ and $t = \frac{2}{3}$ respectively. At these two points on the curve, the Bernstein polynomials $B_1(t) = 3t(1-t)^2$ and $B_2(t) = 3t^2(1-t)$ have maximum values. For the first approximation of the control points, Masood and Ejaz, use chord-length parameterization to find the position on the vector points v which corresponds to approximately the $t = 1/3$ and $t = 2/3$ point on the Bézier curve. They then find the distance between the vector points and the chord joining the curve endpoints at these two points. Masood and Ejaz, then place the intermediary control points at twice the distance between the vector points and the chord, as illustrated in Figure 2.2. To refine the position of these control points, Masood and Ejaz, first estimate the fit error as $\delta(i) = \{(v_i - C) : \min_{c \in C}(\text{abs}(v_i - c))\}$ where C is the Bézier curve, v are the vector points obtained from the drawing. This fit error is then used to determine the displacement required to shift the control points to a better location as:

$$E_k = \frac{\sum_{i=1}^{n} \delta(i) B_k(i)}{\sum_{i=1}^{n} B_k(i)} \qquad (2.16)$$

$k = [1, 2]$ such that the control point is updated using $P'_k = P_k + E_k$.

2.3 Evaluation of Vectorization Algorithms

Quantification of the performance of vectorization algorithms is essential to allow for the comparison of different algorithms. It is also important that algorithms are evaluated with the same evaluation metrics in order to facilitate such comparisons. Different performance metrics have been proposed in literature. In addition to suitable metrics, it is also important to have a benchmark dataset against with algorithms may be compared, thus ensuring that the algorithms can be compared on a like-by-like basis. Here, we describe some performance metrics and a benchmark dataset which evaluate specific properties of the vectorization.

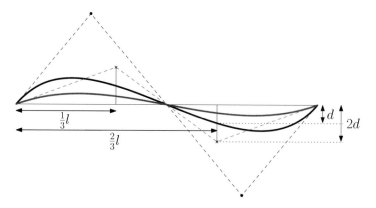

Figure 2.2   An initial estimation of the position of the control points. Control points are placed at a distance of $1/3$ and $2/3$ of the length of the chord formed by the two endpoints of the curve. The estimated control points are placed at twice the distance between the vector points and the chord at these two points. The initial estimation of the Bézier curve and its enclosing polygon are shown in red.

2.3.1 Spatial Accuracy of Junction Points

Since the location of junction points is an important aspect of the vectorization, Noris et al. [33] suggests the use of the salient point error (SPE) defined as:

$$SPE = \frac{1}{|J_g|} \left(\sum_{i \in J_g} \min_{j \in J_v}\{\mathbf{D}(i,j)\} + \sum_{j \in J_v} \min_{j \in J_g}\{\mathbf{D}(i,j)\} \right) \qquad (2.17)$$

where J_g is the set of ground truth junction points, J_v is the set of junction points detected by the vectorization algorithm and $\mathbf{D}(i,j)$ is the Euclidean distance between the detected and ground truth junction points [33]. If the vectorization algorithm locates all the junctions in the image, then $|J_g| = |J_v|$ and the SPE measures the displacement between the detected and ground truth junction positions. Equation 2.17 is normalised by $2|J_g|$ to account for the distance between detected and ground-truth junctions being measured twice for each junction pair. If, however, the vectorization under-segments or over-segments the drawing, then $|J_g| \neq |J_v|$ and the SPE metric is further penalised by the mismatch between the ground truth and detected junctions.

2.3.2 Scalability

The vectorization algorithm should ideally be invariant to scale changes in the drawing and the junction points should be detected equally well across multiple scales [36, 54]. Pham et al. suggest measuring this scale invariance by systematically reducing the size of the image by scaling factors $s_0 > s_1 > , \cdots , > s_i$ through image interpolation. A repeatability score, defined as:

$$R(s_i, \epsilon) = \frac{|\kappa(J_{s_0}, J_{s_i}, \epsilon)|}{\max\{|J_{s_0}|, |J_{s_i}|\}} \tag{2.18}$$

where J_{s_0} is the set of junctions found at the original image scale s_0, J_{s_i} is the set of junctions found at the interpolated scale s_i, $\kappa(J_{s_0}, J_{s_i}, \epsilon)$ is the set of junctions at scale s_i located within a Euclidean distance ϵ of the junctions at J_{s_0}.

2.3.3 Accuracy of the Vectors

Wenyin and Dori [53] define a protocol for evaluating the quality of the complete vector which comprises of two parts, namely, an assessment of the vectors at pixel level and an assessment at a vector level. At pixel level, the metric measures the degree of overlap between the detected line strokes and the ground truth and is defined as:

$$PRI = \alpha D_p + (1 - \alpha)(1 - F_p) \tag{2.19}$$

$$D_p = \frac{|P_g \cap P_d|}{|P_g|}$$

$$F_p = 1 - \frac{|P_g \cap P_d|}{|P_d|}$$

where D_p is the pixel detection rate, F_p is the pixel false alarm rate, P_g the set of ground-truth pixels and P_d the set of detected pixels. The weight parameter $0 \leq \alpha \leq 1$ gives the relative importance of the detection rate in comparison to the importance to the false alarm rate.

At vector level, the metric measures the quality of the detected vector in terms of matching shape, end-point distance and degree of overlap which are represented by their geometric mean Q_v. The vector recovery index is then defined as:

$$VRI = \beta D_v + (1 - \beta)(1 - F_v) \tag{2.20}$$

$$D_v = \frac{\sum_{g \in V_g} Q_v(g)l(g)}{\sum_{g \in V_g} l(g)} \tag{2.21}$$

$$F_v = \frac{\sum_{k \in V_d}(1 - Q_v(k))l(k)}{\sum_{k \in V_d} l(k)} \tag{2.22}$$

where $Q_v(g)$ is the vector quality of the ground-truth vector g with length $l(g)$, $Q_v(k)$ is the vector quality of the detected vector k with length $l(k)$, D_v is the vector detection rate and F_v is the vector false alarm rate, while the weight parameter $0 \le \beta \le 1$ gives the relative importance of the vector detection rate in comparison to the vector false alarm rate.

The two metrics can be combined into a single, combined recovery index $CRI = \gamma PRI + (1 - \gamma)VRI$, where $0 \le \gamma \le 1$.

2.3.4 A Benchmark Dataset

Supervised learning approaches require training images, that is pairs of sketches and their corresponding simplified sketches. In [43] these were generated by asking sketching artists to draw rough sketches over the neat drawing, thus ensuring that the rough sketch and the neat drawing are aligned. There is a major issue with the collection of sketched data in this manner and this is that the roughness of the sketch becomes limited by the underlying neat drawing. Such approaches for data collection would result in trained networks which would then be difficult to generalise for rough sketches found "in the wild" [55]. While Simo-Serra et al. [44] address this issue in a subsequent development of their work with the introduction of the adversarial network which allows for unsupervised training. Nevertheless, the need of a good benchmark data set should be obvious as it allows comparison across different methods, irrespective of the algorithmic approach used. Indeed, a benchmark dataset for "in the wild" sketches is needed to avoid the unintentional bias that may be introduced when a dataset is curated specifically for the assessment of a specific algorithm. Such a dataset should ideally reflect real sketches from sketch-artists collections, drawn without any preconceptions of what the sketch should look like and using the tools and materials typically used by the artist. Unfortunately, while such sketches can be collected relatively easily, these sketches would lack the ground truth counterparts required to use for reliable training or evaluation.

Yan et al. [55] published, what, to our knowledge, is the first benchmark[1] of its kind. A total of 281 sketches all in raster format were collected from 56 different designers. Of these sketches, a subset of 101 sketches were curated for ground-truthing. Ground-truth raster images were created by seven artists hired specifically for the purpose. These artists used Adobe Illustrator to draw over the sketched images and were instructed to make the best guesses to resolve ambiguities while taking care to not add detail not present in the original image. One of the artists was then tasked to create a manual vectorization based on the clean data. Since the original sketches had additional decorative strokes in the form of shading, textures, scaffold lines and text annotations, these were labelled and stored separately. Thus, the dataset consists of the rough images, their simplified and vector representations. This dataset is, therefore, ideal for the evaluation of both simplification algorithms which retain results in raster-format and vectorization algorithms which require clean sketches as inputs and allows for a fair comparison of different approaches.

2.4 Conclusion

This chapter presented an overview of the different facets of the paper-based sketch interpretation problem, starting with sketch simplification and vectorization up to the 3D reconstruction of the sketched object. The inherent ambiguity associated with paper-based sketches as well as the availability and improved technologies of digital stylus and touch-screens saw a decline over the interest in paper-based interpretation systems, in favour for online systems that could provide users with better drawing flexibility owing to the additional data acquired by the system which allowed for some degree of disambiguation of user strokes. The increased interest in deep learning approaches as well as the accessibility of computer hardware for training and testing such deep networks has brought about a renewed interest in the interpretation of paper-based sketches, leading to more works which explore different degrees of drawing ambiguities and drawing styles both at the sketch simplification steps as well as at the 2D interpretation stages. This move has necessitated the availability of large datasets of sketched data exposing a shortfall in the paper-based sketch interpretation community, that is, the lack of a source of curated rough sketches, and their corresponding

---

[1]https://github.com/Nauhcnay/A-Benchmark-for-Rough-Sketch-Cleanup

ground truth data for the community to use in order to compare results. In Section 2.3.4 we describe how Yan et al. [55] have addressed this issue for the simplification of sketches by providing "in the wild" sketches and their corresponding ground-truths. Using this dataset for newer sketch simplification and vectorization works will increase the comparability among different works.

References

[1] P. B. Alvarado-Velazco and V. Ayala-Ramirez. Polygonal Approximation Of Digital Curves Using Genetic Algorithms. In *2012 IEEE International Conference on Industrial Technology*, pages 254–259, 2012.

[2] A. Bartolo, K. P. Camilleri, S. G. Fabri, J. C. Borg, and P. J. Farrugia. Scribbles To Vectors: Preparation Of Scribble Drawings For Cad Interpretation. In *Proceedings of the 4th Eurographics Workshop on Sketch-Based Interfaces and Modeling*, SBIM '07, page 123–130, 2007.

[3] M. Bessmeltsev and J. Solomon. Vectorization Of Line Drawings Via Polyvector Fields. *ACM Transactions on Graphics*, 38(1), January 2019.

[4] T. H. Betaouaf, E. Decencière, and A. Bessaid. Automatic Biometric Verification Algorithm Based on the Bifurcation Points and Crossovers of the Retinal Vasculature Branches. *International Journal of Biomedical Engineering and Technology*, 32(1):66–82, 2020.

[5] L. Boatto, V. Consorti, M. Del Buono, S. Di Zenzo, V. Eramo, A. Esposito, F. Melcarne, M. Meucci, A. Morelli, M. Mosciatti, S. Scarci, and M. Tucci. An Interpretation System for Land Register Maps. *Computer*, 25(7):25–33, 1992.

[6] A. Bonnici, D. Bugeja, and G. Azzopardi. Vectorization of Sketches with Shadows and Shading using COSFIRE Filters. In Proceedings of the ACM Symposium on Document Engineering 2018.

[7] J. Chen, Q. Lei, Y. W. Miao, and Q. S. Peng. Vectorization Of Line Drawing Image Based On Junction Analysis. *Science China Information Sciences*, 58(7):1–14, 2015.

[8] J. Y. Chiang. A New Approach for Binary Line Image Vectorization. In *1995 IEEE International Conference on Systems, Man and Cybernetics.*

Intelligent Systems for the 21st Century, volume 2, pages 1489–1494 vol.2, 1995.

[9] D. Das. A Minutia Detection Approach from Direct Gray-Scale Fingerprint Image Using Hit-or-Miss Transformation. In Asit Kumar Das, Janmenjoy Nayak, Bighnaraj Naik, Soumen Kumar Pati, and Danilo Pelusi, editors, *Computational Intelligence in Pattern Recognition*, pages 195–206, 2020.

[10] I. Debled-Rennesson, S. Tabbone, and L. Wendling. Fast Polygonal Approximation of Digital Curves. In *International Conference on Pattern Recognition*, volume 2, pages 465–468, 2004.

[11] C. Di Ruberto and A. Morgera. A New Algorithm for Polygonal Approximation Based on Ant Colony Optimization. In Pasquale Foggia, Carlo Sansone, and Mario Vento, editors, *Image Analysis and Processing – ICIAP 2009*, pages 633–641, 2009.

[12] S. Di Zenzo, L. Cinque, and S. Levialdi. Run-based Algorithms For Binary Image Analysis And Processing. *IEEE Transactions on Pattern Analysis and Machine Intelligence*, 18(1):83–89, 1996.

[13] L. Donati, S. Cesano, and A. Prati. A Complete Hand-drawn Sketch Vectorization Framework. *Multimed Tools and Applications*, 78: 19083–19113, 2019.

[14] D. Dori and Wenyin Liu. Sparse Pixel Vectorization: An Algorithm And Its Performance Evaluation. *IEEE Transactions on Pattern Analysis and Machine Intelligence*, 21(3):202–215, 1999.

[15] A. A. El-Harby and G. F. Meyer. Automatic Line Extraction By The Square Scan Algorithm. In *Image Processing And Its Applications, 1999. Seventh International Conference on (Conf. Publ. No. 465)*, volume 2, pages 832–835 vol.2, 1999.

[16] D. Elliman. A Really Useful Vectorization Algorithm. GREC '99, page 19–27, Berlin, Heidelberg, 1999. Springer-Verlag. ISBN 3540412220.

[17] J. D. Favreau, F. Lafarge, and A. Bousseau. Fidelity vs. Simplicity: A Global Approach To Line Drawing Vectorization. *ACM Transactions on Graphics*, 35(4), July 2016.

[18] R. F. C. Guerreiro and P. M. Q. Aguiar. Connectivity-enforcing Hough Transform For The Robust Extraction Of Line Segments. *IEEE Transactions on Image Processing*, 21(12):4819–4829, 2012.

[19] X. Hilaire and K. Tombre.   Robust And Accurate Vectorization Of Line Drawings. *IEEE Transactions on Pattern Analysis and Machine Intelligence*, 28(6):890–904, 2006.

[20] R. D. T. Janssen and A. M. Vossepoel. Adaptive Vectorization Of Line Drawing Images. *Computer Vision and Image Understanding*, 65(1):38 – 56, 1997.

[21] Xiaoyi Jiang and Horst Bunke. Fast segmentation of range images into planar regions by scan line grouping. *Machine Vision and Applications*, 7(2):115–122, 1994.

[22] J. Jimenez and J. L. Navalon. Some Experiments in Image Vectorization. *IBM Journal of Research and Development*, 26(6):724–734, 1982.

[23] D. Keysers and T. M. Breuel.   Optimal Line And Arc Detection On Run-length Representations. In Wenyin Liu and Josep Lladós, editors, *Graphics Recognition. Ten Years Review and Future Perspectives*, pages 369–380, 2006.

[24] P. Kruizinga and N. Petkov. Nonlinear Operator For Oriented Texture. *IEEE Transactions on Image Processing*, 8(10):1395–1407, 1999.

[25] J. K. Lakshmi and M. Punithavalli. A Survey On Skeletons In Digital Image Processing. In *2009 International Conference on Digital Image Processing*, pages 260–269, 2009.

[26] R. D. Lins, S. J. Simske, and R. B. Bernardino.   DocEng'2020 Time-Quality Competition on Binarizing Photographed Documents. In *Proceedings of the ACM Symposium on Document Engineering 2020*, DocEng '20.

[27] C. Liu, E. Rosales, and A. Sheffer.   StrokeAggregator: Consolidating Raw Sketches Into Artist-intended Curve Drawings. *ACM Transactions on Graphics*, 37(4), July 2018. ISSN 0730-0301.

[28] X. Liu, T. T. Wong, and P. A. Heng.   Closure-aware Sketch Simplification. *ACM Transactions on Graphics*, 34(6), October 2015.

[29] D. G. Lowe.   Three-dimensional Object Recognition from Single Two-dimensional Images. *Artificial Intelligence*, 31(3):355–395, 1987.

[30] A. Masood and S. Ejaz.   An Efficient Algorithm for Robust Curve Fitting Using Cubic Bézier Curves.   In De-Shuang Huang, Xiang Zhang, Carlos Alberto Reyes García, and Lei Zhang, editors, *Advanced Intelligent Computing Theories and Applications. With Aspects of Artificial Intelligence*, pages 255–262, 2010.

[31] G. Monagan and M. Roosli. Appropriate Base Representation Using a Run Graph. In *Proceedings of 2nd International Conference on Document Analysis and Recognition (ICDAR '93)*, pages 623–626, 1993.

[32] M. Nidelea and A. M. Alexei. Method Of The Square — A New Algorithm For Image Vectorization. In *2012 9th International Conference on Communications (COMM)*, pages 115–118, 2012.

[33] G. Noris, A. Hornung, R. W. Sumner, M. Simmons, and M. Gross. Topology-driven Vectorization Of Clean Line Drawings. *ACM Transactions on Graphics*, 32(1), February 2013.

[34] C. F. Olson. Constrained Hough Transforms For Curve Detection. *Computer Vision and Image Understanding*, 73(3):329 – 345, 1999.

[35] S. Pal, P. Ganguly, and P. K. Biswas. Cubic Bézier Approximation of a Digitized Curve. *Pattern Recognition*, 40(10):2730–2741, 2007.

[36] T. A. Pham, M. Delalandre, S. Barrat, and J. Y. Ramel. Accurate Junction Detection And Characterization In Line-drawing Images. *Pattern Recognition*, 47(1):282–295, 2014.

[37] J. Y. Ramel, N. Vincent, and H. Emptoz. A Coarse Vectorization As An Initial Representation For The Understanding Of Line Drawing Images. In Karl Tombre and Atul K. Chhabra, editors, *Graphics Recognition Algorithms and Systems*, pages 48–57, 1998.

[38] M. Röösli and G. Monagan. Adding Geometric Constraints To The Vectorization Of Line Drawings. In Rangachar Kasturi and Karl Tombre, editors, *Graphics Recognition Methods and Applications*, pages 49–56, 1996.

[39] P. L. Rosin and G. A. W. West. Segmentation of Edges into Lines and Arcs. *Image and Vision Computing*, 7(2):109–114, 1989.

[40] K. Sasaki, S. Iizuka, E. Simo-Serra, and H. Ishikawa. Joint Gap Detection And Inpainting Of Line Drawings. In *Proceedings of the IEEE Conference on Computer Vision and Pattern Recognition (CVPR)*, July 2017.

[41] P. J. Schneider. *An Algorithm for Automatically Fitting Digitized Curves*, chapter XI.8, pages 612–626. Academic Press Professional, Inc., USA, 1990. ISBN 0122861695.

[42] L. Shao and H. Zhou. Curve Fitting with Bézier Cubics. *Graphical Models and Image Processing*, 58(3):223–232, 1996.

[43] E. Simo-Serra, S. Iizuka, K. Sasaki, and H. Ishikawa. Learning To Simplify: Fully Convolutional Networks For Rough Sketch Cleanup. *ACM Transactions on Graphics*, 35(4), July 2016.

[44] E. Simo-Serra, S. Iizuka, and H. Ishikawa. Mastering sketching: Adversarial Augmentation for Structured Prediction. *ACM Transactions & Graphics*, 37(1), January 2018.

[45] S. Simske and M. Vans. *Functional Applications of Text Analytics Systems*, chapter Clustering, Classification and Categorisation, pages 87–133. River Publishers, 2021.

[46] J. Sklansky and V. Gonzalez. Fast Polygonal Approximation of Digitized Curves. *Pattern Recognition*, 12(5):327–331, 1980.

[47] J. Song and M. R. Lyu. A Hough Transform Based Line Recognition Method Using Both Parameter Space And Image Space. *The Journal of the Pattern Recognition Society*, 28:539 – 552, 2005.

[48] J. Song, F. Su, C. L. Tai, and C. Shijie. An Object-oriented Progressive-simplification-based Vectorization System For Engineering Drawings: Model, Algorithm, And Performance. *IEEE Transactions on Pattern Analysis and Machine Intelligence*, 24(8):1048–1060, 2002.

[49] S. Tanigawa, O. Hori, and S. Shimotsuji. Precise Line Detection from an Engineering Drawing Using a Figure Fitting Method Based on Contours and Skeletons. In *Proceedings of the 12th IAPR International Conference on Pattern Recognition, Vol. 3 - Conference C: Signal Processing*, volume 2, pages 356–360, 1994.

[50] K. Tombre, C. Ah-Soon, P. Dosch, G. Masini, and S. Tabbone. Stable and Robust Vectorization: How to Make the Right Choices. In Atul K. Chhabra and Dov Dori, editors, *Graphics Recognition Recent Advances*, pages 3–18, 2000.

[51] Yanglin Tu, Zengwei Yao, Jiao Xu, Yilin Liu, and Zhe Zhang. Fingerprint restoration using cubic bezier curve. *BMC Bioinformatics*, 21(21):514, 2020.

[52] J. Vince. *Mathematics for Computer Graphics*, chapter Curves and Patches, pages 233–257. Springer-Verlag London, 2006.

[53] L. Wenyin and D. Dori. A Protocol For Performance Evaluation Of Line Detection Algorithms. *Machine Vision and Applications*, 9(5):240–250, 1997.

[54] G. S. Xia, J. Delon, and Y. Gousseau. Accurate Junction Detection And Characterization in Natural Images. *International Journal of Computer Vision*, 106(1):31–56, 2014.

[55] C. Yan, D. Vanderhaeghe, and Y. Gingold. A Benchmark for Rough Sketch Cleanup. *ACM Transactions on Graphics*, 39(6), November 2020.

3

Online Interpretation of Sketched Drawings

T. Metin Sezgin

Koç University, Turkey

This chapter will discuss techniques used to sample and acquire the sketched strokes, online sketch recognition, as well as observations on sketching habits, for example, the order with which specific shapes are drawn, which facilitates the interpretation of the sketches. The chapter will also discuss the requirements of interactive sketch-based interfaces, namely the need to make the sketch-based interface as intuitive as possible, without overburdening the user with interactions which interrupt the sketching process. Throughout this discussion, reference will be made to existing, online sketch-based modeling interfaces, noting in particular, how these address the needs and requirements of such an interface.

In offline interpretation, there is plenty of work focusing on 2D drawings depicting 3D shapes. However, much of the work in the online domain has been developed for 2D symbolic sketches. The largest online sketch databases are also of this nature [23, 12, 24]. Hence, we will discuss relevant representation schemes, databases, and algorithms from this bodywork even if they do not directly address 3D object recognition.

As discussed in Chapter 2, in the sketch-based modeling and sketch recognition literature, the word "online" has been used to refer to cases where the timing information is available. In the most general sense, an online sketch is one where the drawing order of the individual strokes is known. More specific definitions may require knowing the beginning/ending timestamp for the individual strokes, and even the timestamps of the individual points making up the stroke as sampled by digitizing hardware. Online algorithms, then, are defined as algorithms that use the timing information for recognizing or processing sketches. Examples of these include temporal [35, 34, 32] and spatio-temporal [5] sketch recognition algorithms, online stroke fragmentation algorithms [43], corner detectors [40, 19, 37], and online active learners [49].

Often, online sketches and online interpretation have been confused with interactive or real-time sketch-based systems. An interactive system that displays recognition results on a continuous basis does not necessarily have to use stroke orderings, or timing information. It could be employing an offline algorithm. The converse is true as well: an online algorithm could be running in the background in a sketch-based interface even if the user interface (UI) itself does not show any apparent signs of interpretation. Similarly, although most offline interpretation algorithms are computationally costly, there are real-time offline algorithms even if their accuracy is not on par with their online counterparts.

The process for offline sketch interpretation usually follows a three-step pipeline:

1. Edge or junction extraction
2. Label generation
3. 3D model reconstruction

Edge and junction extraction have been described in Chapter 2 while the label generation and 3D model reconstruction steps will be described in Chapter 4. Although more recent methods based on deep networks may skip some of these steps in favor of end-to-end processing, the pipeline, and the main sub-processes of online interpretation mimic the offline case. Hence, this chapter starts with three sections describing each process in sequence. In doing so, we compare and contrast the main issues, solutions, and the adopted terminology. Note, that in sketch interpretation, edge and junction detection is framed in the context of primitive extraction and this terminology is adopted in this chapter.

Any online sketch can be rasterized to obtain an offline version of it. Therefore, all algorithms capable of dealing with offline sketches are also adequate for online sketches. Algorithms for offline and online sketches only diverge if, in the latter case, they exploit temporal information associated with the strokes. Hence, in this chapter, we will cover algorithms that use timing information in the primitive extraction, label generation, and interpretation stages. The biggest contrast is seen in the primitive extraction step since timing information associated with the individual points in a stroke allows extraction of the pen speed, which in turn has been shown to be useful for corner detection. Once the primitives are extracted, the label generation and 3D construction steps are typically carried out with the algorithms described in Chapter 4. This is because the timing information is of little value for the remaining steps. An exception is recognition of symbolic 2D objects, and scenes consisting of such objects. In this case, stroke orderings provide substantial temporal cues that aid object recognition and scene segmentation. Hence, algorithms for online stroke processing (Section 3.1) are relevant for 2D as well as 3D sketching, whereas the discussion on sketched symbol recognition (Section 3.2.1) and sketch scene recognition (Section 3.2.2) is primarily relevant for 2D sketches.

3.1 Online Stroke Processing

Online sketches are typically captured using a digitizing tablet setup. We define an online sketch as a sequence of time-ordered strokes $S = S_1, S_2, ...S_N$. Each stroke, in turn, consists of ordered points. Both the strokes, and the individual points may be accompanied with timestamps indicating the time of sampling by the digitizing hardware.

The digitization hardware and the accompanying software used for recording strokes can substantially affect the quality of position and timing information. For example, Figure 3.1 shows a simple square drawn using a Wacom[1] tablet on Windows. As seen here, even this relatively clean drawn square has noisy curvature and speed profiles, mostly caused by digitization error.

Digitization error originates from two sources. First, the spatial resolution of digitizers is limited. Hence, pen location is digitized to a virtual discrete grid. The direction of the pen movement computed over the grid contains digitization error. Curvature of the stroke, defined as the derivative of the direction with respect to the curve length is further corrupted since the curve length is also computed over the same discrete grid. The speed data, defined as the time derivative of curve length, requires using the timestamp of the individual points in the stroke. Since the timestamps of the points are subject to digitization error, this also corrupts the speed information. The operating system and application programming interface (API) related issues further complicate matters by, for example, corrupting the timestamps by poor buffering or occasionally dropping or duplicating points.

Depending on the ink capture setup, it may be possible to collect position information at the native resolution of the digitizer, which is often at a sub-pixel level, and timing information directly through the system clock.

---

[1] https://www.wacom.com/en-us

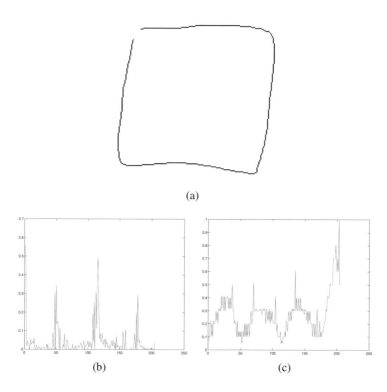

(a)

(b)   (c)

Figure 3.1   A square drawn in a single stroke (a). Its curvature (b) and speed graphs (c). Note how the pen speed drops at the high curvature regions.

In any case, characterization of the target software and hardware platform is recommended.

As its offline counterpart, online stroke processing aims to reduce the input complexity by converting sequence points to meaningful sets of geometric primitives. In the end, each stroke yields one or more time-ordered primitives.

The exact set of output primitives and the process that the stroke undergoes depends on the use case. For example, for some applications, detecting the corners of a stroke are sufficient. Methods geared to address this problem are called corner detectors [37]. While the algorithms described in Section 2.2 may be used for this purpose, for online sketch interpretation, time-quality performance becomes a critical factor. For strokes consisting strictly of line segments, one of the most efficient and practical corner detectors are the Ramer–Douglas–Peucker algorithm, also known as the Douglas–Peucker algorithm [11, 26]. The algorithm is conceptually simple, and very easy to implement. Unlike the optimal curve finder algorithm Sklansky and Gonzalez [38] discussed earlier (Section 2.2, Figure 2.5), the Douglas-Peucker algorithm does not use a cone-like distance metric. Rather, it recursively subdivides the stroke. In each recursive call, it declares the point farthest away from the straight line connecting the endpoints stroke as a corner. The stroke is divided at this point and the same procedure is applied recursively to find further corner points until the orthogonal distance of the farthest point drops below a pre-specified threshold ϵ. The split-and-merge concept used in the Douglas-Peucker algorithm is, therefore, similar to that later adopted by Rosin and West [27] (refer to Section 2.2) but does not assess the goodness-of-fit to measure the validity of the approximation. The Douglas–Peucker is, therefore, faster for instances when the stroke consists of straight lines with no self-intersections. See Figure 3.2 for an illustration of the Douglas–Peucker algorithm.

Another simple algorithm for corner detection for polylines is the ShortStraw algorithm by [47]. This algorithm works by sliding a straw that spans a predetermined number of points along the stroke and calculates the farthest point within the window of interest to the straw line.

The main limitation of the Douglas-Peucker and ShortStraw algorithms is their inability to deal with strokes with curved regions. In such cases, both algorithms may inadvertently declare points on smooth curves as corners. Sezgin's algorithm offers a solution that takes curvature as well as speed information along the stroke to pick true corners [36]. This algorithm exploits the fact that the kinematics of the motor movements required for drawing leads to slower pen movement at and around corners (see Figure 3.1). Corners are extracted by finding minima of pen speed, and the maxima of curvature. In practice, using speed and curvature data is not as straightforward as it may appear.

Later work by Stahovich [40] follows a similar approach improved with a split and merge strategy for corner detection using speed and curvature information. An improved version of the ShortStraw algorithm, IStraw by Xiong and LaViola Jr. [48], detect corner candidates that fall on curved regions by a simple angle test. Both of Stahovich's and Sezgin et al.'s algorithms use speed information as an additional feature. Unlike them, IStraw does not use speed information.

As discussed earlier, and demonstrated in Figure 3.1, both position information and timing information is subject to digitisation errors. The

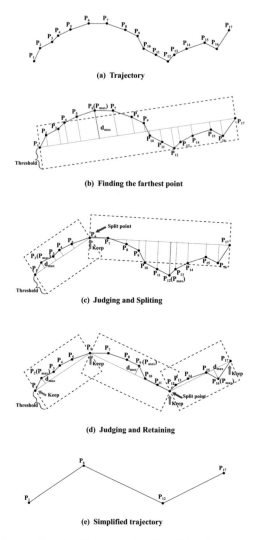

(a) Trajectory

(b) Finding the farthest point

(c) Judging and Spliting

(d) Judging and Retaining

(e) Simplified trajectory

Figure 3.2 The Ramer–Douglas–Peucker algorithm identifying corners by iteratively finding the farthest points [51].

corner candidates obtained by thresholding the speed and curvature data may contain many false corners, which can be eliminated by postprocessing the list of corners Sezgin et al. [36], or by applying a low pass filter. However, picking the right threshold value and filter settings is subject to trial and error, and is a limitation of these relatively simple and easy-to-implement methods.

Once the corners are detected, Bézier curves, arcs, elliptical segments, or other parametric curves can be fit to these regions as described in Section 2.2. Detecting curved regions can easily be accomplished by checking the ratio of the stroke length integrated between successive corners to the euclidean distance between them. This ratio should be near 1 for non-curvature regions.

All methods for corner detection described so far, including those that combine speed and curvature information, depend on hand-tuned thresholds. Hence they generalize poorly across data sets with different noise characteristics. This inability to adapt to noise creates a real problem because different data collection and digitization setups have greatly varying noise characteristics. Sezgin and Davis [33] applied the scale-space idea pioneered by Witkin [46] to the sketch domain.

Figure 3.3 depicts freehand stroke along with the scale space for its curvature data. As seen in this figure, when the curvature signal is convolved with successively wider smoothing filters, the noise gradually disappears. At the extreme, when the signal undergoes no smoothing, all points where the curvature satisfies the local minima properties are detected as a corner. At the other end, where the signal undergoes extreme smoothing, actual corners disappears. Hence we need a method for setting a smoothing factor (scale) that filters out noise without throwing away actual corner points. Sezgin and Davis [33] shows a method for optimal scale selection by keeping track of the number of corner points detected as a function of scale and finding a trade-off between choosing a fine-scale where data is too noisy and introduces many false positives and a coarser scale where true feature points are filtered out. Sezgin and Davis, show that a steep drop in the number of detected corner points is typically detected at the appropriate scale due to the filtering out of the false positives.

A unique property of sketches, which is unlike photographs, is that they are inherently ambiguous, and their meaning is subjective. This ambiguity persists at the stroke level as well. In the context of corner detection, a point that is perceived and declared as a corner may very well be appraised as a non-corner by another person, or in another context (see Figure 3.4). The same goes for the linear versus curved distinction as well. Such subjective appraisal can only be learned from humans by asking them to describe how they perceive the stroke. None of the methods described so far is capable of incorporating human input into account in stroke processing. DPFrag by Tumen and Sezgin [43] is a trainable stroke fragmentation algorithm that learns perceptual preferences from labeled data. Hence, it is capable of

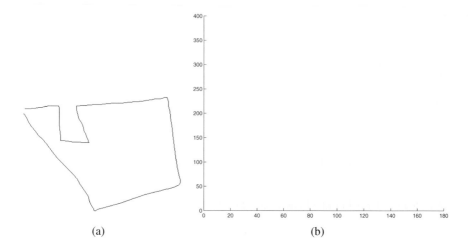

(a)   (b)

Figure 3.3   A freehand stroke (a), and the corresponding scale-space displaying maxima of curvature (b). The vertical axis depicts the smoothing factor, and the horizontal axis shows the index of the points in the stroke where the curvature is locally maximum. A dot in the scale space signifies that the point with the corresponding index had maximum curvature at the given scale.

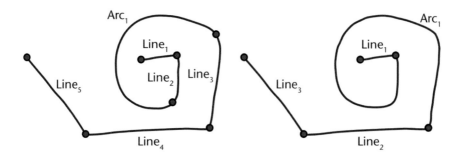

Figure 3.4   Two valid sets of corners are marked for the same stroke. Whether or not the additional points marked as a corner on the left is driven by how one perceives two additional lines ($Line_2$ and $Line_3$) when looking at the stroke as a whole.

generalizing to the unique needs of databases, drawing styles, applications. DPFrag is also unique in that it brings a holistic approach to processing the input stroke using dynamic programming that allows obtaining the optimal segmentation of the stroke.

3.2 Online Sketch Processing

The work on the processing online sketches is generally referred to as "online sketch recognition" in the literature. However, it is essential to make a distinction between the work that focuses on object recognition and scene recognition. Object recognition, also known as isolated object recognition, symbol recognition or gesture recognition assumes the input is that of a single object, and the class is unknown. These methods can be put into use if the sketch-based user interface has been designed in a way that identifies a collection of strokes as being part of the same objects. Scene recognition, also known as full sketch recognition, or sketch scene recognition assumes the input stroke collection may comprise strokes from two or more objects. Interpretation of scenes requires segmentation of the input into groups of strokes, each representing unique objects or gestures.

3.2.1 Sketched Symbol Recognition

Work on recognition of single stroke gestures can be considered as the earliest example of sketched symbol recognition. In this respect, Rubine's gesture recognizer is the simplest and best-known algorithm [28]. Rubine first computes a set of features from an online stroke, for example, speed, duration, and curvature, and trains a linear classifier. The classifier is fast, and accurate as long as the set of gestures is designed carefully to yield linearly separable features.

Among the features suggested by Rubine is the summation of orientation changes along with the gesture. This bears resemblance to the Freeman chain-code which, rather than maintain a record of all angular changes, quantizes these into eight main directions. Assuming a set of exemplar gestures, symbol recognition can then be carried out by comparing the resulting chain-codes using, for example, the minimum edit distance [10].

Most intuitive, however naively simplistic, approaches to multi-stroke sketched symbol recognition are knowledge-based. These approaches treat each object as being comprised of primitives with certain spatial and geometric relations. Recognition proceeds by first dividing user strokes into primitives as described in the online stroke processing in Section 3.1. Then, a rule-based approach is used to check the constraints between the strokes [3]. This approach has the advantage of not requiring any training data, as many later approaches do. However, designing rules specific to each class is an arduous task which is a major disadvantage. Later work by Hammond and Davis [18] brings in the idea of automatically generating the list of primitives

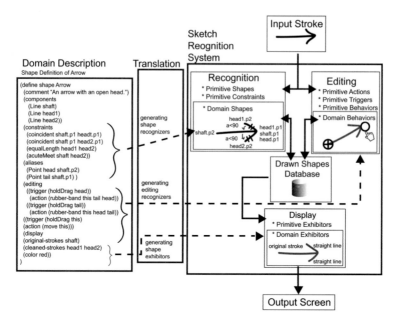

Figure 3.5   An overview of how shapes are defined using the LADDER language and used to automatically construct sketch recognizers [17].

and constraints defining objects from a single prototypical drawn instance. The object definitions are then used to automatically construct recognisers (see Figure 3.5).

The success of knowledge-based approaches has remained limited due to the ambiguous nature of sketches. In particular, any errors committed at the primitive extraction stage propagate to and inhibit the higher-level reasoning process. A solution to this problem is to establish channels of communication between the low-level processing, and higher-level interpretation to facilitate top-down and bottom-up information flow. Alvarado [2] uses probabilistic graphical models to define a probability distribution over the extracted primitives and their combinations.

The object recognition methods listed so far use temporal information in a rather simplistic way, mainly to limit the search space generated from the set of primitives comprising a scene. However, sketching is highly stylized, and there is more to the drawing order than meets the eye. In particular, there is a strong tendency for people to use prototypical drawing order as they lay out the strokes of an object. For example, when asked to draw a school bus, people almost invariably draw the body of the bus, followed by the wheels.

Decorations such as the windows are added last, usually in a left-to-right order. These sorts of regularities have been recognized and studied within the psychology literature, and explained by phenomena such as anchoring, motor convenience, planning, and hierarchy [44, 39, 30]. Sezgin [29], Sezgin and Davis [31], and Sezgin and Davis [32] proposed an approach to capture the statistics of the stroke orderings used for objects to build sketch recognizers. This approach learns the temporal ordering patterns in features and relative properties of the strokes from training data using hidden Markov models (HMMs). The learned statistical models are then used for recognition. This approach pioneered many variations that use HMMs in various ways. For example, Jiang and Sun [22] and Sun et al. [41] propose methods for selecting HMM hyperparameters, while Yuan et al. [50] combine HMMs and support vector machines (SVMs) for classification.

3.2.2 Sketch Scene Recognition

Sketch scene recognition takes a collection of strokes describing a scene with at least two objects. The output consists of a proper grouping of the strokes into sets depicting individual objects along with their names. The grouping operation is referred to as "segmentation", and naming is called "recognition". Segmentation and recognition are often thought of as separate processes, usually, recognition preceding segmentation in greedy algorithms. However, in reality, greedy recognition does not yield accurate segmentation, and segmentation needs recognition results to guide its search process. Hence, there is a chicken and egg problem: perfect isolated recognizers could in theory, solve the segmentation problem, and if segmentation could be solved perfectly, then isolated object recognizers, which are highly accurate, could be used for recognition. Therefore, in practice segmentation and recognition go hand in hand, enumerating, evaluating, and eliminating possible hypotheses together.

As noted earlier, all algorithms capable of dealing with offline sketches are also adequate for online sketches. Hence, here we focus on algorithms that specifically use temporal information for recognition and segmentation, specifically, those that use temporal stroke features and stroke orderings to improve the efficiency of the segmentation process.

The most basic online scene recognition algorithms make use of the available temporal information in a broad sense, by exploiting only the stroke orderings. As noted by [30], during online sketching people typically draw one object at a time. They move on to drawing subsequent objects after they finish those started earlier. Hence, the strokes from each unique object form temporally contiguous sequences in time. This is also called uninterspersed drawing. Gennari et al. [13] propose a greedy approach to a recognition that assumes uninterspersed drawing. Their algorithm works in five steps:

1. ink segmentation,
2. enumeration of candidate symbols,
3. symbol recognition,
4. pruning,
5. error correction.

The first step amounts to segmenting the strokes in the scene into a sequence of primitives as discussed in Section 3.1. In the second and third steps, starting with the first primitive, recognition hypotheses are generated for temporally contiguous sub-sequences of primitives in the scene while keeping upper and lower bounds on the number of primitives allowed for each object type. In the fourth step, these candidate symbols are pruned using heuristics. The last step aims to fix any recognition errors that may have arisen. In addition to typical greedy algorithms that only rely on a measure of confidence in the match, this algorithm uses a set of heuristics to compute a score capturing geometric and contextual consistency. Due to its greedy nature, this method is not guaranteed to find an optimal segmentation. Furthermore, hand-crafting object and domain-specific heuristics renders the method impractical.

Work by Sezgin and Davis [32] presents a joint segmentation and recognition framework based on dynamic programming. The approach first trains isolated object recoginzers, and combines the matching scores they produce over subparts of the scene. The subparts are chosen with the constraint that all the strokes are temporally contiguous. In practice, there are exponentially many partitions of a set of n strokes, however, the temporal contiguity constraint brings this to a number linear in the number of strokes. The scores computed over these groups are brought together in a dynamic programming framework to compute the optimal segmentation in $\mathcal{O}(nc^2)$ time, where n is the number of strokes, and c is the number of distinct object classes. In this work, the object classifiers are based on HMMs, and they use temporal information for object recognition. Therefore, this approach uses online information for object recognition as well as segmentation. In later work, Arandjelović and Sezgin [6] demonstrate that image-based classifiers can be incorporated in place of, or in addition to, temporal classifiers. More notably, it demonstrates that the information of how an object looks (image data) and how it was drawn (temporal data) can be merged to build spatio-temporal scene recognizers. Furthermore, it shows that both of these representations carry unique bits of information that complement each other to yield higher recognition accuracy compared to their unimodal counterparts.

3.3 Requirements of Interactive Sketch-Based Interfaces

Designers, and the end-users of systems that use sketch recognition usually have distinct desires, goals, and expectations. The end-user wants to accomplish a certain task with the technology, while the designer wants to create, evaluate and improve an application [20]. Likewise, the end-users and the designers have different concerns. For example, the end-users are concerned with usefulness, desirability, and usability [20]. They emphasize system reliability, predictability, and minimal distractions [45]. The designers are concerned with designing a system that delivers these by combining the right interaction techniques, the proper look and feel, and the appropriate means of providing feedback [20].

A set of best practices for the design and implementation of sketch-based interface helps strike the middle ground between the end-user and designer concerns, which are often at odds. Various examples of best practices have been reported by researchers since the earlier days of the field. Alvarado [4] formally coined the term SkRUIs to refer to Sketch Recognition User Interfaces, and for the first time shared a set of guidelines for their design and development. Later work was added to these guidelines. In this section, we will elaborate on them.

Sketch-based interfaces have been motivated by the ubiquity, ease of use, and intuitiveness of sketching on paper. They promise to bring the naturalness of paper with the computational power of a computer. Hence, they will achieve their promised capacity to the degree they are able to transfer the desired affordances provided by pen and paper with those of a computer. The following is a list of features that collectively define the design principles for sketch recognition interfaces:

Intuitive and approachable: The term walk-up-and-use has been used to define interfaces that require little or no training to use. In that sense, sketch recognition user interfacess (SkRUIs) should feel intuitive. This is usually achieved by moving the pen and paper metaphor to the UI in terms of the kinds of actions, gestures, and look and feel.

Modeless and button-free: A piece of paper is modeless. During the sketching activity, the user explicitly switches modes by picking a pencil or an eraser. The user also implicitly navigates between creating and consuming the sketch content during the sketching activity. There is no need to switch to note-taking, selection, or drawing mode. Handwriting, symbols, objects and icons can be drawn without changing the pen. This gives the paper and pen setup a tremendous efficiency advantage. Users would ideally like to enjoy this feature in SkRUIs as well, however, it is worth noting that it is a challenging feat from a recognition point of view. Most SkRUIs resort to buttons or toolbars to address a challenge, however desirable they may be. In cases where buttons, or toolbars are unavoidable, they should be designed to be large to facilitate easy targeting [4].

Memorable: The set of gestures, icons, and shortcuts to use should be designed such that they are easily recalled. Although not strictly a sketch-based interface, the difficulties that the Palm Pilot experienced in practice can be attributed to the set of gestures (for example, [28]) that is used, which were hard to learn and remember.

Supporting ambiguity: Sketching nurtures creativity because the ambiguity inherent in rough sketches triggers emergent interpretations [14]. Supporting creative interpretations is particularly essential in early design [8, 21]. Hence, SkRUIs should be designed in a way that supports ambiguous intentions [15]. This is usually achieved by leaving parts of the sketch in raw sketchy form even if it has been recognized.

Supportive of ambiguities: As much as supporting ambiguity matters, there are occasionally cases where the users would like a SkRUI to commit to a particular interpretation.

Transparent: The general UI design principle of transparency advocates designing systems such that relevant aspects of the system state is transparently conveyed to the user.

Consistent feedback: To allow effective communication of the system state, the feedback provided to the user should be consistent and meaningful.

Fluid: The interaction should combine recognition, editing, and drawing in a seamless way such that the flow of thought is not put to halt by mode switches or elements of feedback [4, 7].

Localized menus: Although buttons are generally discouraged effective use of pie menus and their variants can be integrated into SkRUIs. However, to avoid unnecessary traversal of long distances, keep these menus localized [16].

Responsive: Digital ink should not lag behind the stylus during sketching, no matter how fast the stylus moves. Recognition and feedback cues should also be designed to preserve the flow of sketching.

Timely feedback: Earlier work in the field suggested displaying recognition results when the user is done sketching. However, predictive technologies such as sketch-auto-completion [1, 42] have shown to be an effective means of providing recognition feedback during drawing [9].

The list above is by no means exhaustive. Some principles push in contradictory directions. Designing effective sketch recognition interfaces will require a careful study of the domain, the target user profile, limits of the recognition technology, and judicious application of these principles.

3.4 Summary and Conclusions

In this chapter, we summarised algorithms for dealing with sketches where timing information is available. Online sketches have three striking features that make them useful in sketch recognition beyond the raster of pixels available in offline sketches. First, online sketches capture the kinematics of drawing, which in the form of pen speed allows one to infer corners of a stroke. Second, they capture stroke ordering information, which allows efficient recognition and segmentation by exploiting temporal preferences rooted in the psychology of sketching and gestalt psychology. Third, online information in pen speed and object orderings capture *how sketches are constructed*, just as raster information captures *what objects look like*. As such, they provide information beyond what offline methods can supply.

Hence combining online and offline information actually makes higher recognition rates achievable.

In this chapter, we also summarised guidelines for designing sketch recognition user interfaces (SkRUI). Although SkRUIs do not necessarily need to use online recognition algorithms, discussing design principles for SkRUIs in this chapter is fit, since interactive recognition tends to be associated with online methods. The guidelines should not be taken as strict rules to follow. Rather, they should be treated as complementing counterparts of traditional UI and usability principles that have for decades guided the Human-computer interaction (HCI) community [25].

References

[1] Ozan Can Altıok, Kemal Yesilbek, and Metin Sezgin. What auto completion tells us about sketch recognition. In *Proceedings of the Workshop on Sketch-Based Interfaces and Modeling*, 05 2016.

[2] C. Alvarado. Multi-domain Hierarchical Free-Sketch Recognition Using Graphical Models. In *Sketch-based Interfaces and Modeling*, pages 19–54. Springer London, 2011.

[3] C. Alvarado and R. Davis. Resolving Ambiguities to Create a Natural Computer-Based Sketching Environment. In *ACM SIGGRAPH 2007 Courses*, SIGGRAPH '07, page 16–es, 2007.

[4] C. Alvarado. Sketch recognition user interfaces: Guidelines for design and development. pages 8–14, 04 2004.

[5] R. Arandjelović and T. M. Sezgin. Sketch Recognition By Fusion Of Temporal And Image-based Features. *Pattern Recognition*, 44(6): 1225–1234, 2011.

[6] R. Arandjelović and T. M. Sezgin. Sketch Recognition by Fusion of Temporal and Image-Based Features. *Pattern Recognition*, 44(6): 1225–1234, 2011.

[7] J. Arvo and K. Novins. Fluid sketches: Continuous Recognition and Morphing of Simple Hand-drawn Shapes. In *Proceedings of the 13th Annual ACM Symposium on User Interface Software and Technology*, UIST '00, page 73–80, 2000.

[8] Q. Bao, D. Faas, and M. Yang. Interplay of Sketching & Prototyping in Early Stage Product Design. *International Journal of Design Creativity and Innovation*, 6(3-4):146–168, 2018.

[9] A. Blessing, T. Sezgin, R. Arandjelovi, and P. Robinson. A Multimodal Interface for Road Design. Workshop on Sketch Recognition, International Conference on Intelligent User Interfaces, Vol. 20. 2009.

[10] A. Coyette, S. Schimke, J. Vanderdonckt, and C. Vielhauer. Trainable Sketch Recognizer for Graphical User Interface Design. In Cécilia Baranauskas, Philippe Palanque, Julio Abascal, and Simone Diniz Junqueira Barbosa, editors, *Human-Computer Interaction – INTERACT 2007*, pages 124–135, 2007.

[11] D. H. Douglas and T. K. Peucker. Algorithms for the Reduction of the Number of Points Required to Represent a Digitized Line or its Caricature. *Cartographica: The International Journal for Geographic Information and Geovisualization*, 10(2):112–122, 1973.

[12] M. Eitz, R. Richter, T. Boubekeur, K. Hildebrand, and M. Alexa. Sketch-Based Shape Retrieval. *ACM Transactions on Graphics*, 31(4): 1–10, 2012.

[13] L. Gennari, L. B. Kara, T. F. Stahovich, and K. Shimada. Combining Geometry And Domain Knowledge To Interpret Hand-drawn Diagrams. *Computers & Graphics*, 29(4):547–562, 2005.

[14] V. Goel. *Sketches of Thought: A Study of the Role of Sketching in Design Problem-Solving and Its Implications for the Computational Theory of the Mind*. PhD thesis, 1991.

[15] M. D. Gross, and E. Y. Do. Ambiguous Intentions: A Paper-like Interface for Creative Design. In *Proceedings of the 9th Annual ACM Symposium on User Interface Software and Technology*, UIST '96, page 183–192, 1996.

[16] T. Grossman, P. Baudisch, and K. Hinckley. Handle Flags: Efficient and Flexible Selections for Inking Applications. In *Proceedings of Graphics Interface 2009*, GI '09, page 167–174, 2009.

[17] T. Hammond and R. Davis. LADDER, A Sketching Language for User Interface Developers. *Computers & Graphics*, 29-4:518–532, 2005.

[18] T. Hammond and R. Davis. Creating the Perception-Based LADDER Sketch Recognition Language. In *Proceedings of the 8th ACM Conference on Designing Interactive Systems (DIS)*, pages 141–150, 2010.

[19] J. Herold and T. F. Stahovich. SpeedSeg: A Technique For Segmenting Pen Strokes Using Pen Speed. *Computers & Graphics*, 35(2): 250–264, 2011.

[20] J. Hong, J. Landay, A. Long, and J. Mankoff. Sketch Recognizers from the End-user's, the Designer's, and the Programmer's Perspective. 2002. Sketch Understanding, Papers from the 2002 AAAI Spring Symposium. Vol. 2. 2002.

[21] M. Hua. The Roles of Sketching in Supporting Creative Design. *The Design Journal*, 22(6):895–904, 2019.

[22] W. Jiang and Z. X. Sun. HMM-based On-line Multi-stroke Sketch Recognition. In *2005 International Conference on Machine Learning and Cybernetics*, volume 7, pages 4564–4570 Vol. 7, 2005.

[23] J. Jongejan, H. Rowley, T. Kawashima, J. Kim, and N. Fox-Gieg. The Quick, Draw! - A.I. Experiment. https://quickdraw.withgoogle.com/.

[24] R. Niels, D. J. M. Willems, and L. Vuurpijl. The NicIcon Database of Handwritten Icons. In *Proceedings of the 1st International Conference on Frontiers in Handwriting Recognition*, pages 296–301, 2008.

[25] D. A. Norman. *The Design of Everyday Things*. Basic Books, Inc., USA, 2002. ISBN 9780465067107.

[26] U. Ramer. An Iterative Procedure for the Polygonal Approximation of Plane Curves. *Computer Graphics and Image Processing*, 1(3): 244–256, 1972.

[27] P. L. Rosin and G. A. W. West. Segmentation of Edges into Lines and Arcs. *Image and Vision Computing*, 7(2):109–114, 1989.

[28] D. Rubine. Specifying Gestures by Example. In *Proceedings of the 18th Annual Conference on Computer Graphics and Interactive Techniques*, SIGGRAPH '91, page 329–337, 1991.

[29] T. M. Sezgin. Generic And HMM Based Approaches To Freehand Sketch Recognition. *CSAIL Student Oxygen Workshop*, 01 2003.

[30] T. M. Sezgin. *Sketch Interpretation Using Multiscale Stochastic Models of Temporal Patterns*. PhD thesis, Massachusetts Institute of Technology, May 2006.

[31] T. M. Sezgin and R. Davis. Early Sketch Processing with Application in HMM Based Sketch Recognition. In *MIT Technical Report, MIT-CSAIL-TR-2004-053*. MIT, 2004.

[32] T. M. Sezgin and R. Davis. HMM-based efficient sketch recognition. In Robert St. Amant, John Riedl, and Anthony Jameson, editors, *Proceedings of the 10th International Conference on Intelligent User Interfaces IUI'05*, pages 281–283, 2005.

[33] T. M. Sezgin and R. Davis. Scale-Space Based Feature Point Detection for Digital Ink. In *ACM SIGGRAPH 2006 Courses*, SIGGRAPH '06, page 29–es, 2006.

[34] T. M. Sezgin and R. Davis. Temporal Sketch Recognition in Interspersed Drawings. In Michiel van de Panne and Eric Saund, editors, *SBIM '07: Proceedings of the 4th Eurographics workshop on Sketch-based interfaces and modeling*, pages 15–22. Eurographics Association, 2007.

[35] T. M. Sezgin and R. Davis. Sketch Recognition In Interspersed Drawings Using Time-based Graphical Models. *Computers & Graphics*, 32(5):500–510, 2008.

[36] T. M. Sezgin, T. Stahovich, and R. Davis. Sketch Based Interfaces: Early Processing for Sketch Understanding. In *PUI '01: Proceedings of the 2001 Workshop on Perceptive User Interfaces*, 2001.

[37] T. M. Sezgin, T. Stahovich, and R. Davis. Sketch Based Interfaces: Early Processing For Sketch Understanding. In *Proceedings of the 2001 Workshop on Perceptive User Interfaces*, page 22. ACM Press, 2006.

[38] J. Sklansky and V. Gonzalez. Fast Polygonal Approximation of Digitized Curves. *Pattern Recognition*, 12(5):327–331, 1980.

[39] P. Sommers. *Drawing and Cognition: Descriptive and Experimental Studies of Graphic Production Processes*. Cambridge University Press, 1984.

[40] T. F. Stahovich. Segmentation of Pen Strokes Using Pen Speed. In *AAAI Technical Report (6)*, volume FS-04-06, pages 152–158. AAAI Press, 2004.

[41] Z. Sun, W. Jiang, and J. Sun. Adaptive Online Multi-stroke Sketch Recognition Based on Hidden Markov Model. In Daniel S. Yeung, Zhi-Qiang Liu, Xi-Zhao Wang, and Hong Yan, editors, *Advances in Machine Learning and Cybernetics*, pages 948–957, 2006.

[42] C. Tirkaz, B. Yanikoglu, and T. M. Sezgin. Sketched Symbol Recognition with Auto-completion. *Pattern Recognition*, 45(11): 3926–3937, 2012.

[43] R. S. Tumen and T. M. Sezgin. DPFrag: Trainable Stroke Fragmentation Based on Dynamic Programming. *IEEE Computer Graphics and Applications*, 33(05):59–67, 2013.

[44] B. G. Tversky. What Does Drawing Reveal about Thinking? *Invited talk at First International Workshop on Visual and Spatial Reasoning in Design, Cambridge, MA.*, 1999.

[45] P. Wais, A. Wolin, and C. Alvarado. Designing a Sketch Recognition Front-end: User Perception of Interface Elements. In *Proceedings of the 4th Eurographics Workshop on Sketch-Based Interfaces and Modeling*, SBIM '07, page 99–106, 2007.

[46] A. P. Witkin. Scale-Space Filtering. In *Proceedings of the Eighth International Joint Conference on Artificial Intelligence - Volume 2*, IJCAI'83, page 1019–1022, 1983.

[47] A. Wolin, B. Eoff, and T. Hammond. ShortStraw: A Simple and Effective Corner Finder for Polylines. In *Proceedings of the Fifth Eurographics Conference on Sketch-Based Interfaces and Modeling*, SBM'08, page 33–40, 2008.

[48] Y. Xiong and J. J. LaViola Jr. A ShortStraw-Based Algorithm For Corner Finding in Sketch-Based Interfaces. *Computers & Graphics*, 34 (5):513–527, 2010.

[49] E. Yanik and T. M. Sezgin. Active Learning for Sketch Recognition. *Computers & Graphics*, 52:93–105, 2015.

[50] Z. Yuan, H. Pan, and L. Zhang. A Novel Pen-Based Flowchart Recognition System for Programming Teaching. In Elvis Wai Chung

Leung, Fu Lee Wang, Langfang Miao, Jianmin Zhao, and Jifeng He, editors, *Advances in Blended Learning*, pages 55–64, 2008.

[51] L. Zhao and G. Shi. A Method for Simplifying Ship Trajectory Based on Improved Douglas–Peucker Algorithm. *Ocean Engineering*, 166:37–46, 2018.

4

Reconstruction from Sketched Drawings

Alexandra Bonnici and Kenneth P. Camilleri

University of Malta, Malta

The earliest works on the interpretation and reconstruction of drawings as 3D objects concern the interpretation of polyhedral objects, that is, drawings depicting solid bodies which are bound by a finite number of planar faces. The line drawing is, therefore, a projection of the polyhedron onto the picture plane. In such drawings, a line segment corresponds to the edges formed by the intersection of two faces while junctions correspond to the vertices of these edges. Each junction, therefore, represents a point which is shared by two or more faces of the polyhedron. The reconstruction problem can then be posed as the problem of determining the equations of the object planes, using the junction points to define a system of linear equations as shown in Figure 4.1. Solving this system of linear equations provides the missing depth coordinate of each junction point.

As discussed in Chapter 1, a drawing, however, does not represent a unique polyhedron. Theoretically, there can be an infinite number of polyhedra that may be represented by the same drawing [17]. This implies that additional constraints are required to determine the unknown depth parameters [25]. Such additional constraints can be obtained through line labeling algorithms which describe the geometric structure of the drawing, labeling edges as concave, convex, or occluding. These labels effectively constrain the depth coordinates of some vertices with respect to neighboring vertices. These additional constraints should provide the required constraints to recreate the drawing as a 3D model.

The process for obtaining a 3D interpretation of a line drawing of a polyhedron drawing can, therefore, be summarised by the following three steps:

Step 1 Begin by extracting all junction points and lines from the drawing. Assuming that all the visible faces and vertices of the polyhedron are

From junction i

    From plane 1:

$$a_1 x_i + b_1 y_i + c_1 + z_i = 0$$

    From plane 2:

$$a_2 x_i + b_2 y_i + c_2 + z_i = 0$$

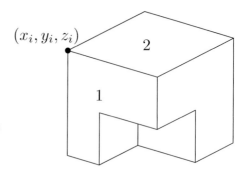

Figure 4.1   A junction is a point shared by two or more faces of the polyhedron. Each junction can be used to form a system of linear equations that describe the planar equation of each object face. In this figure, we illustrate two such equations obtained from the junction with coordinates (x_i, y_i, z_i).

also visible in the drawing, organize the junctions and lines according to the corresponding polyhedral faces.

Step 2 Find consistent labels for the drawing.

Step 3 Assuming that all junctions and lines of the drawing have been located, and consistent labeling can be obtained, find the plane equations to reconstruct the drawing as a 3D model.

The algorithms required for the first step in the interpretation process have been described in Chapter 2 for offline sketches and in Chapter 3 for online sketches. In what follows, we will discuss the state of the art solutions that exist for edge labeling and the drawing reconstruction steps. The discussion will also describe approaches taken to extend the drawing interpretation beyond polyhedron objects and also include objects with curved surfaces and other more free-form shapes.

4.1 Finding Consistent Edge Labels for a Line Drawing

The automatic interpretation of drawing is non-trivial because of the possible drawing errors that may exist in the drawing, particularly when the drawing is sketched in a freehand manner [11]. In Section 1.3 we also discussed that a drawing may have multiple interpretations due to the loss of dimensionality when projecting the 3D form onto the 2D paper plane [17]. Nevertheless, human interpretation of the drawing follows some

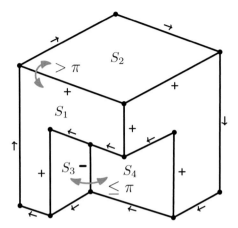

Figure 4.2 The exterior angle between the planes S_1 and S_2 is greater than π and the edge dividing the two planes is labelled as a convex edge using the label **-** . On the other hand, the exterior angle between S_3 and S_4 is less than π and thus, the edge separating the two planes is labelled as a concave edge. Edges on the exterior boundary of the object are occluding a plane and are thus labeled with → edge label.

form of visual constructs which eliminate nonsensical interpretations of the drawing [11]. One approach to encode these visual constructs in the drawing interpretation is through the use of line-labeling.

Line labeling was first introduced by Huffman [12] and Clowes [5] who established a labeling scheme that describes the different ways with which two planes S_1 and S_2 can come together to form an edge. Huffman and Clowes define three edge types, namely convex, concave and occluding edges as shown in Figure 4.2. At convex edges, the exterior angle between the two planes S_1 and S_2 must be greater than π and these edges are denoted by the label **+** . Conversely, at the concave edge, the exterior angle between the two planes S_1 and S_2 is smaller than π and this edge is assigned the label **-** . A concave or convex edge label on the 2D sketch indicates the presence of a visible, physical 3D edge, such that all points lying on that edge are common to the two intersecting planes. Occluding edges occur whenever either one of the planes S_1 or S_2 is not visible to the observer. Unlike concave and convex edges, occluding edges are labeled by the directional label → where, by convention, the direction of the arrow is such that the occluded plane is on the left-hand side of the edge when facing the direction of the arrow [7].

In 2D sketches, occluding edges can occur at edges that apparently separate two visible faces. In such cases, the edge is a feature of only one of the planes, by definition, of that on the right-hand side of the directional edge. This implies that the two planes are disjoint in 3D with the plane on the right-hand side of the directional edge being closer to the observer. The task of an edge-labeling algorithm can then be described as finding the set of edge labels that can be assigned to each edge of the drawing in such a manner that the labeling drawing describes a physical 3D object. One way to ascertain that the labeling scheme selected represents a valid physical object is to determine that the edge interpretations at each junction represent a valid physical junction. It is also important to ensure consistency, that is, the edge interpretation obtained from the junctions on either side of the edge is the same. The most common way to determine if a sketched edge labeling results in a feasible 3D object is to compare the labels at each junction against a junction-label catalog which lists all possible edge labels at each junction.

Note that the three edge labels described above are suitable for simple trihedral objects, that is, objects where, at most, three planes can be present at any given junction. However, more complex objects, including objects containing curved surfaces, require a richer description of the edges. The edge labeling problem can then be described as comprising of three components, namely:

- identifying the junction geometries that completely describe the class of objects being labeled
- determining the edge labels that are possible at these junctions
- obtaining an efficient label assignment mechanism through which the edge labels can be checked such that a suitable edge label interpretation is found for the drawing

A description of these tasks is now given.

4.1.1 Determining the Edge Interpretations

Huffman and Clowes note that trihedral drawings can be described by four junction types, namely the W, L, T, and Y junctions shown in Figure 4.3. Edges formed by planes intersecting at these junctions can be fully described as concave, convex, or occluding. Simple trihedral objects can be grouped together to form more complex objects. Waltz [28] notes that by bringing together trihedral objects, new junction and edge interpretations are formed when the concave or convex edge of one object comes into contact with an occluding edge or a surface of another object. Waltz describes such junctions and edges as the *extended trihedral* geometry, identifying five new junctions, namely the T-L, Y-L, W-L, and W-W junctions shown in Figure 4.3.

In addition to these new junctions, Waltz also introduces the concept of *separable* edges, that is, edges which do not pertain to a single object but which are common to two or more trihedral objects as illustrated in Figure 4.3. Waltz introduces three new edge labels. *Separable concave* edges, with the label $\rightarrow$, occur when an occluding edge of one object rests against a visible surface of a second object such that an apparent concave edge is formed. *Separable convex* edges, with the label $\overset{+}{\rightarrow}$ occur when an occluding edge of one object is aligned with the convex edge of the second object, such an edge forms a seam between the two objects. The *3-object* edge, with the label $\overset{+}{\rightarrow\!\!\!*}$ occurs when two occluding edges of two different objects align with the concave edge of the third object, forming an apparent concave edge. The separable edges are directional, taking the direction of the occluding edge. Moreover, since the separable edges occur on the edge boundaries of more than one object, they provide additional information about the spatial arrangement of the objects in the sketch. When the separable edges are located on the exterior boundary of the foreground object, the separable edges provide information about the spatial relation between the foreground object and the background. On the other hand, when the separable edge occurs within the exterior boundary of the object, they give an indication that the object can be divided into sub-parts and the separable edges are the edges along which the object can be separated into the constituent components.

Varley and Martin [27] note that from a small sample of 56 engineering drawings, around 75% of the drawings have trihedral or extended-trihedral geometries, while the remaining 25% have tetrahedral geometries. Thus, Varley and Martin introduce three new junctions, namely the K, M, and X junctions. These junctions do not require additional edge interpretations and the trihedral edge labels can be used to describe the edges at these new junctions.

Objects may contain curved as well as planar surfaces. In such cases, the geometry that describes the drawing must take into account the instances where the curved surface meets the planar surface, adjusting the trihedral junctions to allow for one or more of the edges at the junction to be a curved edge. In addition, the geometry describing the object must also take into account the projection of a 3D curved surface on the 2D plane since this introduces edges that are viewpoint dependent and not physical edges, that is, the drawn edge does not represent a physical discontinuity on the object surface. Malik describes these edges as *extremal* edges with the label $\twoheadrightarrow$. Just like the occluding edge label, the extremal edge is a directional edge,

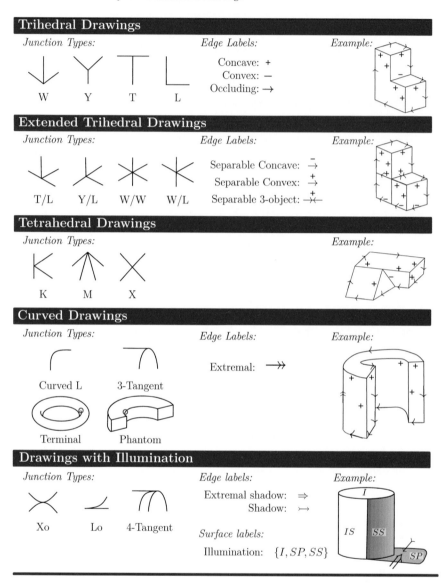

Figure 4.3   The junctions and labels used for different drawing types.

with the direction of the arrow such that the hidden part of the surface is on the left of the edge when facing the direction of the arrow. Moreover, Malik also introduces the 3-tangent junctions where the straight line in the junction represents such an extremal edge.

Malik also introduces two additional junctions, namely the *terminal* junction and the *phantom* junction. Both these junctions are pseudo-junctions, that is, they are not physical junctions but need to be introduced to take into account discontinuities introduced particularly with circular or elliptical surfaces. Terminal junctions occur at the end of the open-ended arc which is drawn to represent the apparent depth discontinuity when drawing elliptical or circular rings [19, 7]. Phantom junctions occur along concave curved surfaces. In such drawings, at the point where the observer's viewpoint is tangential to the curve, the curved surface appears to self-occludes, thus changing the edge interpretation along the length of the edge, as illustrated in Figure 4.3. However, changes in the edge interpretation along the edge are typically associated with inconsistencies in the edge labeling, which, in turn are associated with incorrect drawings. By introducing the phantom junction along such an edge is effectively split into two shorter segments, thereby allowing a change interpretation along the edge [19, 7].

Scene illumination has an important role in the interpretation of a drawing, reinforcing the interpretation of convex or concave shapes as well as the spatial relationship between different objects. For this reason, Waltz [28] and Cooper [7] also considers the labeling of scenes containing shadows, specifically, the labeling of drawings where the shadow boundaries are represented as edges in the drawing. Waltz introduces the shadow label ⟫→ , where the label is placed across the shadow edge with the arrow pointing into the shaded region. Waltz further introduces surface labels that provide an indication on the illumination of each surface. Three labels are used, namely, the illuminated surface IS label which indicates that the surface is fully illuminated and has no shadow, the projected shadow SP label which indicates regions that are in shadow because illumination is obstructed by another object, and the self-shadowed SS label where the surface faces away from the light source. Cooper extends the shadow labels to include the labeling of curved surfaces with shadows. Here the extremal shadow edge, with the label ⟹ is introduced to identify the shadow boundary due to the extremal edges. Cooper also notes that new junctions are formed at the point where the shadow boundary touches the object edge, introducing the 4-tangent, Xo and Lo junctions as shown in Figure 4.3.

4.1.2 Creating Junction Dictionaries

In addition to the edge and junction geometries, the labeling problem requires knowledge in the possible edge labels at any given junction. Such knowledge

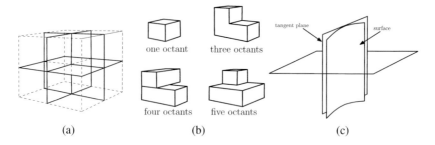

one octant   three octants

four octants   five octants

tangent plane   surface

(a)   (b)   (c)

Figure 4.4 (a)To establish a junction-label catalog, the 3D space is divided into octants [28] (b) By filling different octants with solid objects different objects are created that list all possible junctions and their interpretations. (c) Malik [19] extends this catalog to allow for curved surfaces by using the dividing planes as tangent planes.

allows the labeling algorithm to determine whether the labeling represents a valid 3D geometry. This knowledge can be presented in a junction-label catalog or dictionary. To do so, the 3D space is divided into octants as shown in Figure 4.4(a). Objects of differing complexity can be created by filling different octants with solid material as shown in Figure 4.4(b). This approach exposes all possible junction geometries and their corresponding interpretations, thus forming the junction dictionary, essentially, a look-up table that lists all plausible edge interpretations given a specific junction geometry [28, 27]. Malik adjusts the division of the space to cater for curved surfaces by using the divisions as tangent planes to curved surfaces as shown in Figure 4.4(c). Waltz further places light sources in different octants, thus obtaining the required labels of shadow edges.

4.1.3 Labelling the Sketch

Having established the junction-label dictionary we next need to find the set of labels that consistently describe the 3D geometry of the object. A combinatorial approach that evaluates all possible labels at each junction requires considerable computational effort and is, therefore, unsuitable for sketches of any degree of complexity. Therefore, alternative labeling schemes are necessary to find a solution to the line labeling problem with minimal computational effort. Waltz [28] introduces a back-tracking approach often referred to as *Waltz filtering*, which arbitrarily assigns an edge label to an edge, pruning the possible edge labels of adjacent edges such as only those consistent with the selected edge are retained. If a valid label is determined, the labeling moves on to the subsequent junction. If however, no valid label is found, the search back-tracks to the last valid junction, and alternative solutions are evaluated [28]. To speed up the labeling process, Waltz divides the junction dictionary into two subsets, one involving junctions associated with interior edges and the other involving junctions associated with exterior edges. Junctions located on the exterior boundary of the object have fewer valid interpretations than interior edges and can, therefore, be labeled with a lower computational effort. Once the exterior edges have been labeled correctly, their labels can be propagated onto the interior edges, limiting the number of interpretations of the interior edges. However, not all interior edges can be completely restricted by the exterior edges. For example, the stem and bar of a T junction are typically disjoint in space such that the edge labels selected for the bar do not bear upon the edge labels of the stem and vice versa [28]. Thus back-tracking may still be required on the interior edges.

Variants of the Waltz-filtering exist. For example, Malik first labels all concave edges, collapsing all other labels into a generic label to refine at a second run of the labeling algorithm [19]. Alternatively, Kirousis first labels particular junctions, specifically Y-W chains, before labeling other junctions [14]. Varley and Martin divide the junction dictionary into trihedral and tetrahedral junctions, using the spatial neighborhood of the junction to determine which of the two dictionaries to use, limiting the use of the larger tetrahedral dictionary only to the immediate vicinity of tetrahedral junctions. In these cases, the motivation is to restrict the number of possible label combinations by labeling the most restrictive labels.

These junction labeling schemes use the junction dictionary constraints as hard constraints, and can only label the drawing if labels that are consistent with the junction dictionary are found. While such hard constraints ensure that drawings of impossible 3D objects are identified, they also fail to label inaccurate sketches of objects which can still be interpreted by the human observer, despite the inaccuracies. Moreover, drawings may have multiple interpretations due to the ambiguities in representing 3D objects as 2D drawings. Cook and Agah [6] rightly argue that human observers do not adopt the same all-or-nothing approach to the interpretation of the sketch. Thus, a labeling scheme should allow the sketch to be labeled in the best possible manner even if fully consistent labeling cannot be determined.

To address the issue of multiple plausible interpretations, Cooper introduces soft constraints of planarity and parallelity to restrict, but not completely prohibit the assignment of labels to edges. Such soft constraints guide the interpretation of the drawing towards the most plausible interpretation of the drawing. While this approach addresses the problem of multiple interpretations, it does not consider the problem of accidental inconsistencies and thus, if a drawing fails to satisfy the junction dictionary constraints, it is still considered unlabelable. Myers and Hancock [20] and Bonnici and Camilleri [1] use a genetic algorithm (GA) to determine the best fitting labels in the drawing. The fitness function which guides the evolution of the GA is based on the junction dictionary such that the dictionary acts as soft constraints to the labeling problem. When drawing inconsistencies occur, the GA will not be able to determine a labeling solution with the maximum fitness but will determine the labeling that best fits the drawing.

Although Myers and Hancock [20] allow for the best fitting label in the case of drawing inconsistencies, the algorithm does not distinguish between a drawing interpretation and its Necker inverse. This ambiguity in the interpretation of the sketch is also common among human observers. Drawings will, therefore, often include artistic cues such as shading to reinforce one interpretation over another. Waltz [28] and Cooper [7] use shading as a further constraint on the edge labeling. However, these methods assume that the shading cues are optically correct representations of the shadows that occur in scenes. Just like sketches are drawn inconsistently, sketched shading too can be inconsistent with optical laws. Thus, cues, just like the junction-label dictionary must be used as soft constraints.

Consider a drawing with N_E edges and N_J junctions where the jth junction has a geometry β_j with valence O_j. An edge label λ_i is associated with each edge in the drawing such that $\epsilon_j = \{\lambda_i\}$ is the set of edge labels associated with the junction j. The drawing may be represented by the chromosome $E = [\lambda_1, \cdots, \lambda_{N_E}]$ with each gene representing an edge label. The fitness of the chromosome may be determined by comparing the edge labels against the junction dictionary D_J and can be expressed as:

$$F(E) = 1 - \frac{1}{2N_E} \sum_{j=1}^{N_J} \phi_j \qquad (4.1)$$

$$\phi_j = \min\{\boldsymbol{H}(\epsilon_j, D_J(\beta_j))\} \qquad (4.2)$$

where $\boldsymbol{H}(\cdot)$ is the Hamming distance and $D_J(\beta_j)$ is the junction dictionary entries associated with the junction geometry β_j [20]. The evolutionary mechanisms of mutation and cross-over allow the GA to evolve an initial population guided by the fitness function, effectively exploring the search-space to converge to a solution within this space. Since the drawing may have multiple interpretations, there will be multiple solutions in this space and the GA may oscillate between these solutions.

Bonnici and Camilleri [1] note that the cues present in the drawing constrain the interpretation of the edge to a subset of all possible edge interpretations. Using this observation, a cue-dictionary D_C can be created and this dictionary is used to constrain the initial population of chromosomes such that only interpretations that are consistent with the cues present at the edge are used to form the initial population. This focuses the initial population onto a particular solution in the search space. The evolutionary processes of cross-over and mutation allow this initial population to deviate from this focused space. While this is potentially of benefit as it allows the GA to explore the search space if the cues are inconsistent, with no additional constraints, the GA may drift away from the solution intended by the artistic cues present in the drawing. It is, therefore, necessary to extend the chromosome fitness evaluation such that the chromosomes that adhere to the cue constraints achieve a higher fitness than others that do not. The fitness function is thus redefined as

$$F(E) = 1 - \left(\alpha \frac{1}{2N_E} \sum_{j=1}^{N_J} \phi_j - (1 - \alpha) \frac{1}{N_E} \sum_{i=1}^{N_E} \theta_i \right) \qquad (4.3)$$

$$\phi_j = \min\{\boldsymbol{H}(\epsilon_j, D_J(\beta_j))\} \qquad (4.4)$$

$$\theta_i = \min\{\boldsymbol{H}(\lambda_i, D_C(c_i))\} \qquad (4.5)$$

where $D_C(c_i)$ is the interpretation associated with the cues c_i present at the edge. The parameter α is a weight factor that determines the confidence in the cues present in the drawing. This approach steers the interpretation of the drawing towards the interpretation suggested by the artistic cues present in the drawing, and, in the case of drawing inconsistencies, towards the probable interpretations of the drawing.

4.2 Reconstructing the Drawing

Once the drawing is vectorized and appropriately labeled, the final step of the 3D interpretation of the drawing would be to assign the depth coordinate to each of the identified vertices and edges. The human interpretation of a drawing as a 3D model is governed by visual regularities which govern the typical projection of the real 3D model as a 2D drawing. These regularities may be loosely categorized into three types [11]. Type-1 regularities govern relationships between low-level entities, for example, the projection of parallel edges, the preservation of vertical lines between the drawing plane and the 3D world. Type-2 regularities govern mid-level support, governing the preservation of orthogonal corners or planarity of faces among others. Type-3 regularities offer more holistic support to the drawing interpretation by governing the relations of the projection of the 3D world onto the 2D plane [11].

Consider a drawing that represents the orthographic projection of a polyhedral object with the viewer located at infinity in the positive direction along the z-axis. Type-3 regularities govern the mathematical representation of this projection such that the face j of the polyhedron is defined as:

$$a_j x + b_j y + z + c_j = 0 \qquad (4.6)$$

where (a_j, b_j, c_j) are the parameters that define the plane, (x, y, z) are the coordinates of a point on the plane with (x, y) corresponding to the pixel coordinates in the (x, y) coordinate system of the 2D drawing while z is the unknown depth value. Each plane of the polyhedron drawing is defined by a minimum of four vertices whose (x, y) coordinates are known. Equation 4.6 can then be expressed as a system of linear equations which, when solved, provide the unknown plane and depth parameters. The polyhedron will have more than one face and thus the system of equations needs to be defined and solved for each plane. Moreover, adjacent planes will have shared vertices, thus fixing the z coordinates of the shared vertices.

Let V be the set of junction points and F be the set of faces in the drawing. For each junction point, we may form a set of vertex-face pairs $R = (v, f | v \in V, f \in F)$ such that the vertex v lies on the face f. The triple (V, F, R) is known as the *incidence structure* of the drawing. From this incidence structure, we may form a system of equations

$$A\mathbf{w} = 0 \qquad (4.7)$$

where $\mathbf{w} = [z_1, z_2, \cdots z_n, a_1, b_2, c_2, \cdots a_m, b_m, c_m]^T$, is a vector of unknown parameters, A is a $|R| \times (3m+n)$ constant matrix where each row corresponds to the (x, y) coordinates of the vertices for the vertex $v_j \in R$ in Equation 4.6, $n = |V|$ is the number of vertices, and $m = |F|$ is the number of faces of the polyhedron. The problem of reconstructing the drawing then becomes the mathematical problem of solving the system of linear equations of Equation 4.7.

Theoretically, there can be an infinite number of polyhedra that may be represented by the same drawing [17]. The consequence of this is that the rows of A are not all linearly independent and the number of independent rows can be expressed by the $\text{rank}(A(R))$. Since matrix A will be rank-deficient, then, there will necessarily be z-coordinates that must be specified independently for the drawing to represent a unique polyhedron. Specifically, a line drawing with an incidence structure $I = (V, F, R)$ which has two or more faces, requires that $\text{rank}(A(R)) + 4 \leq n + 3m$ to be reconstructable in this manner [25].

Although some of the unknown z coordinates are specified independently, the values selected should be such that the reconstructed polyhedron represents the geometric structure of the line drawing. The line labeling described in Section 4.1 provides the required geometric structure since the edge labels provide information about the relative position of each face and vertex from the viewer. Using this information, for each vertex i located further away from the viewer than face j, we can form a system of linear inequalities governed by:

$$a_j x_i + b_j y_i + z_i + c_j > 0 \qquad (4.8)$$

These inequalities can be collectively expressed by

$$B\mathbf{w} > 0 \qquad (4.9)$$

where B is a constant matrix where each row represents the coefficients from Equation 4.8. We can, therefore, reconstruct a three-dimensional polyhedron from a drawing by solving the set of equations and inequalities given by Equation 4.7 and Equation 4.9 [25].

However, in practice, drawings lack the accuracy needed for the vectorization process to yield vertices whose coordinates are such that Equation 4.7 and Equation 4.9 have a solution. As discussed in Section 2.1, drawing roughness may cause the detected vertices to drift from their geometrically correct positions which will reduce the number of linearly independent rows in A such that Equation 4.7 becomes inconsistent. In such cases, however, in the absence of any other drawing errors, the inequalities described by Equation 4.9 will still hold. More serious drawing errors can occur in the drawing where vertices shift to such an extent that faces which should not be coplanar appear coplanar. In such cases, the inequalities described by Equation 4.9 will also be inconsistent.

From a human interpretation perspective, it is easy to notice the inconsistency of the drawings that violate the inequalities of Equation 4.9.

However, drawings which cause inconsistency in Equation 4.7 still appear interpretable. Thus, while it is acceptable for 3D interpretation systems not to find solutions to drawings for which Equation 4.9 is inconsistent, 3D interpretations for drawings for which only Equation 4.7 is inconsistent are expected. In 3D reconstruction literature, this problem is known as the *super-strictness problem*. To resolve the super-strictness problem, we need to find a subset of faces $X \subseteq F$ for which we may form a reduced incidence structure $I = (V(X), X, R(X))$ where $V(X)$ is the set of vertices on the faces $X \subseteq F$ and $R(X)$ is the corresponding incidence pairs. From this incidence structure, a reduced, but consistent, system of equations $A^* \mathbf{w} = 0$ is formed and used together with Equation 4.9 to reconstruct the drawing. This reduced system of equations can be achieved by omitting faces, or moving vertices to better locations until a satisfactory system of equations is obtained [23, 24, 21]. Random selection and movement of vertices however can cause significant deviations of the reconstructed object from the design intent. Thus, although Type 3 regularities provide a framework for solving the reconstruction problem, the Type 1 and Type 2 regularities should be taken into consideration to resolve the super-strictness problem in a way that matches the human interpretation of the sketch [22]

Lipson and Shpitalni argue that a threshold-based approach to determining whether a 3D reconstruction complies with the reconstruction regularities is too rigid and unlike the human perception of such regularities. They, therefore, model the adherence to the regularities through a continuous compliance factor $\mu_{a,b} = \exp(-((x-a)/b)^2)$ which compares an observation x to some ideal value a with a standard deviation b. This compliance factor, therefore, returns a value of 1 when the observation is a match for the ideal value and decays to 0 the further the observation departs from the ideal. The regularities introduced by Lipson and Shpitalni [17] include:

Planarity: For a plane, defined by the planar face equation $ax+by+cz+d = 0$, the distance of a point from the place is defined by $|ax + by + cz + d|$. For a point on the plane this absolute difference should be equal 0. This regularity measures the absolute difference of the reconstructed vertices from the plane whose coefficients are obtained through the incidence graph structure.

Line parallelism: Lines that are parallel in the 2D plane should also be parallel in the 3D plane. This regularity measures the degree of parallelity between pairs of lines in the 2D plane and likewise between their reconstructed 3D counterparts.

Line verticality: Vertical lines in the sketch plane should also be vertical in the 3D plane. Similar to the parallelism regularity, this regularity measures the verticality of lines in the sketch plane and the corresponding verticality of the 3D reconstruction of the same line.

Isometry: Lengths of entities in the 3D model should be uniformly proportional to their lengths in the sketch plane. This regularity, therefore, measures the length ratios between corresponding 3D and 2D lines.

Corner orthogonality: W and Y junctions in the sketch plane should be reconstructed as orthogonal corners. This regularity measures the orthogonality between the reconstructed 3D edges at such junctions.

Skewed face orthogonality: A face contour that shows skewed orthogonality in the sketch-plane is probably orthogonal in the 3D space.

Skewed facial symmetry: A face showing skewed symmetry in the sketch-plane denotes a true symmetric face in 3D.

Line orthogonality: All line pairs in a junction that are not collinear are most likely perpendicular. This regularity is used to obtain initial inflation of the drawing.

The minimum standard deviation of angles: Since line strokes are, in their greater part aligned with one of the three main axis of the 3D model, the angles formed between all pairs of lines meeting at junctions should be similar. This regularity is used to favor 3D reconstructions for which the standard deviation between the junction lines is small, implying that the reconstructed edges are aligned with the main axis of the 3D world.

Face perpendicularity: All adjacent faces should be perpendicular. This follows from the line orthogonality regularity above and is also used to obtain initial inflation of the drawing.

Prismatic face: A face joining two parallel arcs in the projection plane is prismatic in space. This regularity is introduced to resolve the reconstruction of cylindrical shapes and the planarity regularity. By dividing elliptical arcs into piece-wise linear segments, cylindrical shapes can be represented as prismatic faces.

Line collinearity: Lines that are collinear in the sketch-plane are also collinear in space.

Planarity of skewed chains: When a chain of edges is found to exhibit skewed symmetry, the 3D reconstruction of the chain should exhibit both planarity and symmetry. Likewise, if a chain of edges exhibits skewed orthogonality, its 3D reconstruction should exhibit planarity and orthogonality.

Assuming that the (x, y) coordinates of each 3D vertex are obtained from the sketch-plane, then Lipson and Shpitalni propose to find the missing z coordinates through an optimization-based approach, casting the optimization a problem in form $F(z) = W^\intercal \sum [\alpha]$ where α is a vector that represents the nine regularities described above, while W is a pre-fixed weight balancing vector. By defining the regularities in the form of continuous, differentiable algebraic terms Lipson and Shpitalni ensure that the optimization process will converge, although the solution found may not necessarily give z coordinates that correspond to the optimal, human-perceived solution. Lipson and Shpitalni compare Brent's minimization, the Conjugate gradient and Genetic Algorithms, reporting that they consider Brent's minimization approach to be the most suitable of the three, since it converges to the correct solution more often and in fewer iterations than the other two approaches.

Lipson and Shpitalni note that an angular distribution graph (ADG) could be used to speed up the optimization. The ADG is a histogram of the orientations of the strokes that are present in the sketch plane. Since the sketch is expected to depict a 3D object with the majority of the edges are aligned with the 3D world axis, such a histogram is expected to have three peaks as illustrated in Figure 4.5, where the orientations of these peaks correspond to the orientations of the main axis. Aligning lines to these orientations, therefore, results in better initial inflation of the drawing.

Kang et al. [13] also use the ADG in their reconstruction algorithm. In their approach, Kang et al. select the vertex whose local angular distribution has the highest correlation with the ADG and anchor this vertex to a depth $z = 0$. A maximum spanning tree is then created, connecting this vertex to all other vertices of the sketch. The weights of this spanning tree are determined as a function of the alignment to the axis system identified through the ADG such that edges which are more closely aligned to the axis directions are assigned a higher weight. In this manner, the maximum spanning tree allows the algorithm to first set the depths of those vertices whose edges are

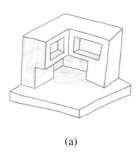

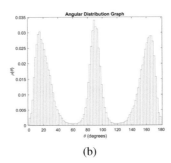

(a)   (b)

Figure 4.5 The angular distribution graph of a trihedral drawing is expected to have three peaks corresponding to the main axis orientations.

aligned with the axis system. Depths are assigned to each vertex based on an optimization approach taking into account the regularities that govern axis alignment, orthogonality and isometry. Kang et al. [13] note that drawings may exhibit multiple axis systems. In such instances, the ADG will exhibit more than three peaks and Kang et al. propose to cater for such drawings by considering each axis system locally.

Tian et al. [26] use orientation similarity to identify those line strokes in the sketch which are outliers, that is, whose orientations differ from the main axis orientations. These line strokes are removed from the initial depth reconstruction such that the initial depth estimation of the drawing is obtained from the more regular lines and vertices. In this algorithm, the driving regularity is the preservation of parallel relationships in the sketch-plane into the 3D reconstruction. Tian et al. show that the parallel relationships between line strokes in 3D can be expressed by a system of linear equations $P\mathbf{x} = 0$ where x is a vector defining the unknown vertex depths and the depth ratios necessary to preserve the parallel relationship between pairs of lines. Tian et al. note that the solution to this system of equations will be the null space of P which can be determined through singular value decomposition (SVD). Thus, P is expressed as $P = U\Lambda V^{\mathsf{T}}$, where U and V are orthogonal matrices and Λ is a diagonal matrix whose values are the singular values of P. The columns of V which correspond to zero singular values form a basis for the null space of P. Since the null space P_{null} is typically smaller than P, this reduces the burden of the optimization step required to determine the unknown depth values.

Due to the sketch roughness, the inaccuracies in the representation of parallel lines in the sketch plane will result in an SVD of P in which Λ has no

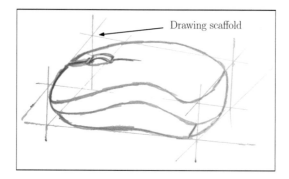

Figure 4.6   Construction scaffolds are used by artists and designers to guide the depicted 3D form of the drawing.

singular value which is exactly zero. To allow for sketch roughness, Tian et al. sets a threshold on the magnitude of the singular values that can be considered sufficiently small to form the null space. Optimization based on the minimum standard deviation of angles is then used to determine the unknown depth values. This approach provides suitable reconstructions of drawings of polyhedral objects governed primarily by parallel surfaces. Although some drawing roughness is allowed, the drawings considered are still relatively neat in comparison to drawings sketched at the concept-level. Gryaditskaya et al. [9] relax the drawing constraints to allow for rougher sketches using the principle of axis-aligned straight lines to introduce the concept of construction scaffolds as illustrated in Figure 4.6. Gryaditskaya et al. argue that many freehand sketches lack the stroke connectivity information that would determine which of the 2D stroke intersections correspond to the artist intended 3D curve intersections. Such inaccuracies will render the creation of the incidence graphs difficult and by mistakenly treating occlusions as 3D intersections will connect distant parts of the shape with consequences on the 3D reconstruction. Gryaditskaya et al. note that despite the sketch roughness and incompleteness, scaffolds are often used by artists and designers to guide the depicted 3D form of the drawing. Regularities of axis alignment, orthogonality and planarity are used together with the regularity of coverage, that is, designers, avoid overshooting lines past the intended endpoints and tangentiality, which takes into account instances when long continuous strokes are over-sketched as shorter segments tangent to the underlying stroke. Axis alignment, orthogonality, planarity, and tangentiality are grouped into a single geometric score and this, together with the coverage score form a total candidate line score which determines the suitability of the 3D interpretation of the candidate line. In their evaluation, Gryaditskaya et al. demonstrate that the suggested reconstructions agree with the human interpretation of the drawing in 78% of the drawings tested with instances of disagreement occurring in drawings which human participants also found ambiguous. The concept of using scaffolds as drawing guides will be seen again Chapter 8 where they are used as drawing guides in 3D sketching.

The methods described thus far assume that the sketch is drawn as a wireframe representation, that is, the sketch has both visible and hidden views of the object sketched out. The hidden views are necessary for the object to be reconstructed in its entirety from a single sketch and in the absence of these hidden views, the full reconstruction of the object requires sketches that show the object from multiple views. Sketching full wireframe drawings or drawing multiple views is not necessarily feasible, particularly for inexperienced users. It is, therefore, desirable to have algorithms that can deduce the hidden views from the visible parts of the drawing, allowing the interpretation algorithm to reconstruct a full 3D model based on informed assumptions about the hidden views [2, 4].

In drawings that depict only the visible faces and vertices, a face can be described as being:

- **fully visible** and consists only of visible vertices
- **completely invisible** and is completely hidden from view
- **partially visible** and has visible edges and vertices but with one, or more vertices being a T-junction, therefore, indicating that the face is partially occluded by another face
- **almost invisible** where the face is framed by L-junctions

as illustrated in Figure 4.7(a). Partial and almost invisible faces have edges and visible vertices which provide clues to the shape of the hidden vertices and hence the hidden part of these faces can be deduced from the drawing. Completely invisible faces, however, consist of only hidden vertices and, with no clue on their shape, cannot be determined [4]. The drawing vertices too can be described according to their visibility having

- **complete** junctions such as the W or Y junctions whose edges are completely visible
- **broken** junctions, notably the T junction where a face occludes another
- **incomplete** junction such as the L-junction where the vertex has unsketched hidden edges.

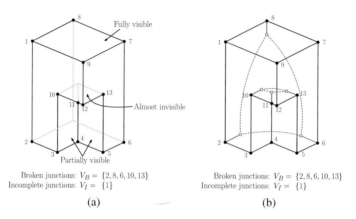

Broken junctions: $V_B = \{2, 8, 6, 10, 13\}$
Incomplete junctions: $V_I = \{1\}$

(a)

Broken junctions: $V_B = \{2, 8, 6, 10, 13\}$
Incomplete junctions: $V_I = \{1\}$

(b)

Figure 4.7   Finding the hidden vertices and junctions in drawings. (a) A drawing will have fully visible faces, almost invisible faces, and partially visible faces. In this example, with five broken junctions, and one incomplete junction, the drawing can have up to six hidden vertices. (b) Each hidden vertex is connected to one incomplete junction. Each hidden vertex is then connected to two other hidden edges and a series of cutting and merging operations are then used to simplify the hidden structure.

Cao et al. determine that the number of hidden vertices N_H in a drawing is $N_H \leq |V_I| + |V_B|$ where V_I is the set of incomplete junctions and V_B is the set of broken junctions.

To determine the hidden structure, Cao et al. start with the maximum possible hidden vertices connecting each incomplete vertex to one, different, hidden vertex. Assuming a trihedral structure, each hidden vertex is then connected to two other hidden vertices, selecting the hidden vertices to connect based on the closeness along the visible boundary of the corresponding incomplete vertices. Cao et al. [2] then proceed to reduce the initial hidden structure to one with fewer hidden vertices through a series of cutting and merging operations, that is, removing an edge and merging two vertices it connects into a single vertex, thus, creating multiple, plausible hidden structures as illustrated in Figure 4.7. The most plausible hidden structure is then selected based on the Gestalt principles of similarity and preferences for symmetry. The 3D reconstruction then proceeds from the reconstructed wireframe drawing.

Rather than first creating the wireframe drawing from the visible sketch to then reconstruct the 3D model from this wireframe, Chansri and Koomsap project the visible faces onto the 3D world to then determine the partially

	Visible Junctions							Broken Junctions	Incomplete Junctions				
	1	9	7	5	3	4	12	11	8	6	2	10	13
1		1							1		1		
9	1		1				1						
7		1							1	1			
5						1				1			1
3						1					1	1	
4				1	1			1					
12		1						1					1
11						1	1					1	
8	1		1										
6			1	1									
2	1				1								
10					1			1					
13				1			1						

Figure 4.8 The vertex connection matrix created for the drawing in Figure 4.7(a). The matrix is divided into nine regions, with the top left region, shaded in blue, representing visible faces, while the bottom right, shaded in gray, represents any invisible face that may exist in the drawing. The middle region, shaded in orange, represents coordinates on partially visible faces.

visible and almost invisible faces through the intersection with the visible faces. To achieve this, Chansri and Koomsap create a vertex-to-vertex matrix marking the connectivity between the visible vertices. By grouping junctions according to the three visibility categories described above, the matrix can be divided into nine regions, according to the face visibility as illustrated in Figure 4.8. Similar to the approach adopted in Kang et al. [13], the ADG is used to determine the drawing's main axis. Likewise, a reference, fully visible vertex is selected and set to depth zero in the 3D model. Then, isometric projections and the three vertices on the visible face are used to determine the face projection from the sketch plane to the 3D plane. By progressively simplifying the vertex-vertex matrix and propagating depth values from the known faces, the depth values of the remaining visible vertices are obtained. The remaining invisible vertices are then obtained from the intersections of adjacent, visible planes.

Both approaches generate the simplest hidden structure based on the clues available through the visible drawing. While this is in agreement with the general viewpoint assumption, more viewpoints are required if the object contains features that are not completely observable from a single view drawing.

Similar to other aspects of the drawing interpretation problem, recent works in the 3D reconstruction of sketches have also started to rely on the use of convolution neural networks, using the network to learn the regularities that govern the 3D inflation through examples learned from different sketch-3D model example pairs. Deep learning approaches have been adopted by [30] to use an example-based retrieval approach to begin the 3D reconstruction process with a semi-ready model, while [15, 18, 10] generate 3D models from multiple sketches depicting the object from different viewpoints.

Lun et al. [18] train a CNN to reconstruct multi-view depth and normal maps from either a single sketch or multiple sketches, with the user able to add new sketches from new viewpoints to refine the reconstruction. The CNN accepts as input $256 \times 256 \times C$ gray-scale image containing the sketched drawings with additional views incorporated as different channels C. Adopting an approach similar to the use of the canonical views in traditional CAD systems, a different network is trained for each input view. The CNN used has an encoder part and a decoder part. The encoder consists of a series of convolutional layers with kernels of size 4 and stride, using batch normalization and a leaky ReLU activation function. The output of the encoder is a $2 \times 2 \times 512$ representation which encodes shape information based on the input sketches. This output can be used for shape retrieval as will be detailed further in Chapter 5. The decoder has 12 separate branches each containing upsampling and convolutional layers, each representing a different viewpoint and which do not share parameters. The output of the decoder is of size $256 \times 256 \times 5$ and consists of a depth-map (1 channel), a normal map (3 channels), and a foreground probability map which gives the probability of foreground vs background and is used as a mask on the dept and normal maps. The convolutional layers in the decoder use a kernel of size 4 with a stride of 1, with each convolutional layer being followed by batch normalization and a leaky ReLU activation function. The output layer uses tanh activation function to normalize the depth and normal maps to the $[-1, 1]$ range and normal maps are further passed through L_2 normalization to ensure that the normals are of unit length.

Training the network requires a considerable amount of 3D shapes and their corresponding sketches. Since curating such a data-set would typically require considerable effort, Lun et al. propose to use a mix of human-generated sketches and sketch-object pairs created through NPR (refer to Chapter 5). To allow the network to learn to map the sketch into the corresponding depth and normal maps, the network is trained with a loss function that penalizes the difference between the ground truth and predicted the depth maps, the angle difference between the ground truth and predicted normals, disagreement between the ground truth and predicted foreground masks, as well as large scale structural differences between the ground truth and predicted maps. The depth and normal maps are then brought together into a single coherent 3D point cloud. Lun et al. note that the predicted depth and normal maps will not necessarily be consistent with each other and, moreover, the derivatives of the predicted depth maps do not necessarily agree with the normals. Thus, Lun et al. propose an optimization approach to reconcile the depth maps and the normals from all viewpoints. The normal maps are used for energy minimization to promote consistency between the depth derivatives and the surface normals. The depth map is used as the z coordinate value and the 3D point cloud is generated using the extrinsic camera projections, that is, the frame rotation and translation in the object space. Each point is assigned a normal based on the predicted normal maps, following this, the iterative closet point algorithm could be applied to rigidly align all point sets. Lun et al. note that such a rigid alignment would result in noisy point clouds and so, they propose to optimize the depth values following three criteria: first that the resulting depth values are close to the predicted depths produced by the network; second, that they are first-order derivatives yield surface tangent vectors that are as orthogonal as possible to the predicted normals and; lastly, that they are consistent with the depth and normal maps of corresponding 3D points generated from other views. Lun et al. then apply the screened Poisson surface reconstruction algorithm to convert the resulting point cloud and normals to a surface mesh. The proposed CNN was trained using 30 human-generated sketches and a further 90 sketches NPR-generated sketches of cartoon characters, airplanes and chairs.

The CNN-based approach is, therefore, a departure from the traditional 3D reconstruction algorithms were described earlier. The output of the CNN is a depth and normal map defined over all the foreground regions of the sketched object, not only the sketched edges. This allows the direct generation of the point cloud and the surface mesh. The same can be said for other CNN-based reconstruction approaches. The problem now shifts towards creating 3D models which capture fine features and are a faithful expression of the sketch, allowing for a variety of sketched objects and at different levels of ambiguity.

Li et al. [15] argue that the sparsity of the sketched drawing would stump a single regression model and, therefore, propose to use a two-stage CNN. The first stage, named the DFNet, regresses a flow-field that is a dense signal which describes the surface curvature of the object. The second stage, named the GeomNet, takes the sketch and the flow-field to predict the depth and normal map and confidence maps. Like Lun et al. [18], the network's input is a 2D sketch as a grayscale image. Li et al. [15] assume that the sketched drawing will be a parallel projection of the 3D object on the sketch-plane and further specify that the object contour strokes should be in pure black while other strokes depicting ridges, valleys or other suggestive contours should have non-zero gray-level values similar to the style described in NPR literature and which are discussed further in Chapter 5. The network used here takes two additional and optional inputs. The first is a depth sample image with manually defined depth values corresponding to specific key points in the sketch, while the second optional input consists of curvature hints in which abrupt changes of the surface normals can be specified. When these optional inputs are provided the regressed depth and surface normals are encouraged to match the specified depth values, allowing for a finer surface rendering. The network was trained and tested on images depicting animals, statues, and characters all of which contained detail that could be represented with reasonably uncluttered 2D sketches. Similar to Lun et al. [18] the sketches are created through NPR. Also similar to Lun et al. [18], this approach allows for sketches from multiple views, using optimization to merge the surfaces due to possible inconsistencies between views.

Han et al. [10] take a different approach, using the CGAN architecture for the reconstruction. The CGAN is used to translate the sketch into an image that reveals the shape geometry from a certain view angle. The predicted images from multiple viewpoints are used as targets for direct shape optimization using a project-and-compare loss to push the reconstructed shape to match the target image from the same view angle. Rather than using a depth map of surface normals, Han et al. use an attenuation image to capture the shape model, that is, an image in which each pixel represents the attenuation along a ray passing through the shape. This attenuation image is learned through the generator network of the CGAN by mapping sketches S_i and a random noise vector z to real attenuation images d_i that is $G : \{s_i, z\} \mapsto d_i'$. The generator G is trained to predict an attenuation image d_i' that cannot be distinguished from the real attenuation image d_i by the discriminator network D which, in turn, is trained to distinguish the real from the predicted attenuation images. Direct shape optimization is then used to obtain the shape model M from the attenuation image. Model M is considered as a voxel grid with resolution $R \times R \times R$. A voxel m in this grid is a variable whose value lies in the range $[0, 1]$ representing the voxel occupancy probability. The shape M can be back-projected to the 2D plane through an orthogonal projection to the desired viewing angle. The scope of the optimization is to then ensure that the shape M appears the same as the attenuation image d_i' predicted by the CGAN from each viewing angle, that is, given the projected image r_i from M the optimization function will minimize the L_2 distance between d_i' and r_i for all viewing angles i.

4.3 Evaluation of Reconstruction Algorithms

The evaluation of the 3D reconstruction algorithms centers on the comparison of the 3D objects created by the proposed algorithms and the ground-truth reconstruction of the object. This evaluation process, therefore, requires some ground-truth 3D models and their corresponding line drawings. There are two schools of thought on how this could be achieved. On the one hand, assuming that we have at hand some line drawings of objects, we may manually create the corresponding 3D model using some CAD tools. This approach was adopted by Xue et al. [30] for 3D models that consisted primarily of planar surfaces and requires a repository of line drawings from which 3D models can be unambiguously created.

A more common approach would be to start with the 3D object and generate the corresponding line drawing from the object. The sketch is then reconstructed back by the algorithm under evaluation, thus obtaining the reconstructed object O_R and the corresponding ground truth object O_G [18, 15, 8, 10]. This approach is more practical since there are various sources from where the 3D object may be obtained. These include public online repositories such as "The Model Resource"[1] used by Lun et al., models generated primarily for segmentation such as the Shape COSEG[2] dataset [29]

---

[1] https://www.models-resource.com/
[2] http://irc.cs.sdu.edu.cn/~yunhai/public_html/ssl/ssd.htm

used by Delanoy et al. and Han et al. as well as datasets generated primarily for semantic labeling such as ShapeNet[3] [3] which is also used by Han et al..

The task then lies with obtaining the sketched line drawings that correspond to the objects. One approach would be to ask human artists to draw the required sketches. This approach would, however, require considerable human effort. Alternatively, the sketches may be generated synthetically. Delanoy et al. render the 3D objects as 2D drawings using a contour rendering approach, applying an edge detector to the normal and depth maps of the object rendered from a specific viewpoint. Similarly, Lun et al. use a combination of NPR techniques including edge-preserving filters on shapes rendered through simple shading schemes and suggestive contours to generate multiple line renderings from the single ground truth 3D object. These multiple line drawings serve as a form of data augmentation for CNN-based approaches. The reader is referred to Chapter 5 for further description of the NPR techniques.

We can now focus on the comparison of the ground truth O_G and reconstructed O_R 3D objects. For planar objects, it is sufficient to compare the relative positions of the object vertexes and the orientations of the edges at these vertexes. These can be measured through two metrics [30]. The first metric is the root mean square of the angle differences defined as:

$$RMSA(O_G, O_R) = \sqrt{\frac{1}{N_a} \sum_{i=1}^{N_a} (O_G^i - O_R^i)^2} \qquad (4.10)$$

where N_a is the number of angles made by any two edges meeting at a vertex and O_G^i, O_R^i are the ith angles from O_G and O_R, respectively. The second metric is the root mean square of the Euclidean distance (RMSE) between the corresponding vertices [30].

The planarity constraint ensures that the 3D object's shape is well defined through the vertices and edges such that the RMSA and the RMSE are sufficient measurements with which to measure the similarity between planar objects. However, in the case of more organic or free-form shapes, the surface shape may differ despite edges appearing to be at the same position. Thus, alternative metrics are required to compare two such objects. These may include:

---

[3]https://shapenet.org/

Hausdorff distance between each surface point of O_R to the surface point on O_G, using the maximum of the distance as a measure of similarity between the rendered surfaces.

The Surface normal difference which is measured as the angular difference between the surface normal at a point in O_R and the corresponding point in O_G.

Depth map errors which are measured as the absolute difference between pixel depths in the depth-maps rendered for O_R and O_G gave a reference viewpoint.

Intersection over Union (IoU) of the volume. Here, both ground truth and reconstructed objects are rendered as volumes using the same voxel resolution. The intersection would then be the number of voxels filled by both the ground truth and the reconstructed objects, while the union would be the total number of filled voxels.

Finally, in CNN-based approaches, one must also take into consideration the division of the dataset into training, validation, and testing sets. While it is common to randomly divide the dataset into these subsets Lun et al. caution against this random division, noting that it may be possible to have near-identical objects, subject to an affine transform in the dataset. In such instances, over-fitting of the model would yield an overly optimistic estimation of the performance of the algorithm. To prevent this, Lun et al. select the testing set based on two selection criteria. The first criterion picks a test shape and aligns this to all remaining shapes in the dataset through the best affine transform. The Chamfer distance from the candidate test object surface to the nearest other objects in the dataset is computed. If this distance is greater than some pre-selected threshold the candidate test object can proceed to the second criterion. This criterion compares the rendered synthetic sketch for the candidate test shape and the closest matching shape in the dataset, computing the Euclidean distance between the corresponding drawings. The object-sketch pair is accepted as a test set if this distance too is above a threshold.

4.4 Conclusion

In this chapter, we note the interpretation of drawings as 3D objects. In our discussion, we note the evolution of the interpretation algorithms from those that are applicable to neat representations of rigid objects to algorithms that offer more support to rougher, concept-like sketches. The more recent CNN-based approaches borrow from the concepts of sketch-based retrieval which are further explored in Chapter 5 to provide a starting point from which 3D shapes may be fine-tuned to the sketched representations.

We note that although there is increased support for concept sketches, algorithms described in the literature mostly ignore the visual cues present in the drawing, focusing mainly on the sketched object contours. While concept sketches can contain a number of drawing cues, of particular interest for the 3D interpretation problem would be the cues associated with shading since these provide additional information about the shape of the object. Shape-from-shading is not a new concept in computer graphics. However, shading in concept sketches may suffer from the same ambiguities and errors that are associated with the contour boundaries. Moreover, sketched shadows differ from shading in scene images since shadows can be sketched in various ways, from smooth shading to hatched regions and even abstracted as shadow contours. Just as deep learning has shown that it is possible to simplify over-traced strokes, a deep learning approach may potentially allow for the simplification of sketched shadow strokes into shadow regions. Likewise, training 3D reconstruction networks with sketches that contain shading may assist in the interpretation of surfaces. Deep learning approaches require suitable data sets for training. Researchers are using datasets created for NPR studies, by reversing their use, that is, passing the NPR sketch through the reconstruction algorithm to achieve the target 3D form. Similar to the observations made in Chapter 2, while such datasets provide for a useful initial starting point for comparison between different algorithms, it would be better to have 3D interpretations of "in the wild" sketches such that the full pipeline from paper to 3D model may be evaluated and compared.

The works presented in this chapter show that different algorithms cater to different sketching styles and varying degrees of sketch ambiguity. This leads us to believe that a single sketch interpretation algorithm with the ability to cater to different drawing styles and interpretation needs is most likely difficult to achieve and users would be better served with the option of selecting the interpretation algorithms best suited to the drawing at hand. Moreover, user participation in user studies show that, while most users agree on an interpretation of a sketch, some user-specific preferences exist. Thus, paper-based sketch interpretation should allow for the possibility of user-based adjustments of the algorithmic outcome. From the user's point of view, such adjustments should be effortless and intuitive. To achieve such goals, the paper-based sketch interpretation would most likely benefit from sketch-based interactions which are described in Chapter 3 and Chapter 8.

References

[1] A. Bonnici and K. P. Camilleri. A Constrained Genetic Algorithm For Line Labelling Of Line Drawings With Shadows And Table-lines. *Computers & Graphics*, 37(5):302–315, 2013.

[2] L. Cao, J. Liu, and X. Tang. What the Back of the Object Looks Like: 3D Reconstruction from Line Drawings without Hidden Lines. *IEEE Transactions on Pattern Analysis and Machine Intelligence*, 30 (3):507–517, 2008.

[3] A. X. Chang, T. Funkhouser, L. Guibas, P. Hanrahan, Q. Huang, Z. Li, S. Savarese, M. Savva, S. Song, H. Su, J. Xiao, L. Yi, and F. Yu. Shapenet: An Information-rich 3D Model Repository. arXiv preprint arXiv: 1512.03012 (2015).

[4] N. Chansri and P. Koomsap. Sketch-based Modeling From A Paper-based Overtraced Freehand Sketch. *The International Journal of Advanced Manufacturing Technology*, 75(5):705–729, 2014.

[5] M. B. Clowes. On Seeing Things. *Artificial Intelligence*, 2(1):79–116, 1971.

[6] M. T. Cook and A. Agah. A Survey Of Sketch-based 3D Modeling Techniques. *Interacting with Computers*, 21(3):201 – 211, 2009.

[7] M. Cooper. A Rich Discrete Labeling Scheme For Line Drawings Of Curved Objects. *IEEE Transactions on Pattern Analysis and Machine Intelligence*, 30(4):741–745, 2008.

[8] J. Delanoy, M. Aubry, P. Isola, A. A. Efros, and A. Bousseau. 3D Sketching Using Multi-View Deep Volumetric Prediction. *Proceedings of the ACM Computer Graphics and Interactive Techniques*, 1(1), 2018.

[9] Y. Gryaditskaya, F. Hähnlein, C. Liu, A. Sheffer, and A. Bousseau. Lifting Freehand Concept Sketches Into 3D. *ACM Transactions on Graphics*, 39(6), November 2020.

[10] Z. Han, B. Ma, Y. S. Liu, and M. Zwicker. Reconstructing 3D Shapes From Multiple Sketches Using Direct Shape Optimization. *IEEE Transactions on Image Processing*, 29:8721–8734, 2020.

[11] D. D. Hoffman. *Visual intelligence: How we create what we see.* WW Norton & Company, 2000.

[12] D. A. Huffman. Impossible Object As Nonsense Sentences. *Machine Intelligence*, 6:295–324, 1971.

[13] D. J. Kang, M. Masry, and H. Lipson. Reconstruction Of A 3D Object From A Main Axis System. In *AAAI fall symposium series: Making pen-based interaction intelligent and natural*, volume 3, 2004.

[14] L. M. Kirousis. Effectively Labeling Planar Projections of Polyhedra. *IEEE Transactions on Pattern Analysis and Machine Intelligence*, 12 (2):123–130, 1990.

[15] C. Li, H. Pan, Y. Liu, X. Tong, A. Sheffer, and W. Wang. Robust Flow-Guided Neural Prediction for Sketch-Based Freeform Surface Modeling. *ACM Transactions on Graphics*, 37(6), December 2018.

[16] H. Lipson and M. Shpitalni. Conceptual Design and Analysis by Sketching. *Artificial Intelligence for Engineering Design, Analysis and Manufacturing*, 14(5):391–401, November 2000. ISSN 0890-0604.

[17] H. Lipson and M. Shpitalni. Optimization-based Reconstruction Of A 3D Object From A Single Freehand Line Drawing. In *ACM SIGGRAPH 2007 Courses*, SIGGRAPH '07, page 45–es, 2007.

[18] Z. Lun, M. Gadelha, E. Kalogerakis, S. Maji, and R. Wang. 3D Shape Reconstruction from Sketches via Multi-view Convolutional Networks. In *2017 International Conference on 3D Vision (3DV)*, pages 67–77, 2017.

[19] J. Malik. Interpreting Line Drawings of Curved Objects. *International Journal of Computer Vision*, 1(1):73–103, 1987.

[20] R. Myers and E. R. Hancock. Genetic Algorithms For Ambiguous Labelling Problems. *Pattern Recognition*, 33(4):685–704, 2000.

[21] L. Ros and F. Thomas. Correcting Polyhedral Projections For Scene Reconstruction. In *Proceedings 2001 ICRA. IEEE International Conference on Robotics and Automation*, volume 2, pages 2126–2133 2001.

[22] L. Ros and F. Thomas. Overcoming Superstrictness In Line Drawing Interpretation. *IEEE Transactions on Pattern Analysis and Machine Intelligence*, 24(4):456–466, 2002.

[23] K. Sugihara. Mathematical Structures of Line Drawings of Polyhedrons—Toward Man-Machine Communication by Means of Line Drawings. *IEEE Transactions on Pattern Analysis and Machine Intelligence*, PAMI-4(5):458–469, 1982.

[24] K. Sugihara. A Necessary and Sufficient Condition for a Picture to Represent a Polyhedral Scene. *IEEE Transactions in Pattern Analysis and Machine Intelligence*, 6(5):578–586, May 1984.

[25] K. Sugihara. *Machine Interpretation of Line Drawings*, volume Artificial Intelligence Vol (1). Cambridge: MIT Press, 1986.

[26] C. Tian, M. Masry, and H. Lipson. Physical Sketching: Reconstruction and Analysis of 3D Objects from Freehand Sketches. *Computer aided design*, 41(3):147–158, 2009.

[27] P. A. C. Varley and R. R. Martin. The Junction Catalogue For Labelling Line Drawings Of Polyhedra With Tetrahedral Vertices. *International Journal of Shape Modeling*, 7(1):23–44, 2001.

[28] D. Waltz. *Understanding Line Drawings of Scenes with Shadows*, chapter 2, pages 19–91. McGraw-Hill, 1975.

[29] Y. Wang, S. Asafi, O. van Kaick, H. Zhang, D. Cohen-Or, and B. Chen. Active Co-Analysis of a Set of Shapes. *ACM Transactions on Graphics*, 31(6), 2012. ISSN 0730-0301.

[30] T. Xue, J. Liu, and X. Tang. Example-based 3D Object Reconstruction from Line Drawings. In *2012 IEEE Conference on Computer Vision and Pattern Recognition*, pages 302–309, 2012.

5

Reconciling Search Results with User Sketches

Paul L. Rosin[1] **and Juncheng Liu**[2]

[1]Cardiff University, Wales, UK
[2]University of Otago, New Zealand

The discussion thus far has focused on the creation of 3D models from 2D sketches, describing offline sketch processing in Chapter 2 and online sketch processing in Chapter 3. These approaches assume that the user is creating a 3D representation of the design concept from scratch, without reference to other designs. However, designers may look for inspiration from existing objects, and a natural form of interaction with other existing objects through sketch-based searches and interactions. This chapter will describe non-photorealistic techniques which can be used to aid these search interaction techniques.

Three-dimensional objects may be rendered as sketched drawings either in model space or image space. The former approach directly analyses the geometry of the 3D object model to extract lines. Alternatively, the latter approach first synthesizes a set of projected images which are generated from a set of views of the 3D model. Two-dimensional image processing techniques are then applied to detect lines in these images. This chapter continues to describe a range of methods for performing sketch-based 3D shape retrieval. The user provides a sketch as input, and matches to it will be searched for in a database of 3D shapes which have been rendered as sketched drawings. Since the 3D shapes have been transformed to look more similar to the input query, this can improve the effectiveness of the score function used to compute match similarity, and thereby facilitate the search. We cover traditional approaches as well as some of the recent deep learning approaches to both line detection and shape retrieval.

We will take a computer vision view of some aspects of sketch retrieval in this chapter. Sketch retrieval shares many elements that are common and extensively researched in the wider area of computer vision, and many techniques that were developed for content based image retrieval (CBIR) and particularly query by example, as well as medical image analysis, remote sensing, etc. can be deployed for sketch retrieval. Image query-based commercial CBIR systems became available in the 1990s (for example, IBM's QBIC – Query By Image Content [17]) and such "reverse image search" services have become popular with mainstream providers such as Google, TinEye, and Bing. Although they generally rely on relatively simple image-based features such as color, texture, and shape, some of them will also use additional (non-image based) metadata when available.

While some of the challenges that were identified in CBIR are mostly specific to object retrieval from images (for example, coping with variations in image resolution, illumination conditions, color, background clutter, multiple objects), others are equally relevant to sketch retrieval.

A significant problem for retrieval, identified by CBIR researchers in the 1990s is the "semantic gap", which is the difference between the low-level features extracted from the data and the manner in which the query is understood by the user in terms of high-level concepts. The use of high-level concept ontologies and high-level features (for example, semantic segmentation) can alleviate this problem by enabling matching to be performed at a high, more abstract, level. But difficulties still exist, especially when the domain of retrieval is broad. It is difficult to both comprehensively capture and represent all the semantics, as well as to extract such high-level features from images.

The number and size of image collections are increasing enormously; for example, many datasets now contain more than a million items. This requires retrieval methods to have sub-linear time performance and low memory footprints. Likewise, sketch datasets are increasing in volume, encouraging similar performance demands. For their convenience, users should not have to enter laborious, detailed queries, and so it should be possible to perform retrieval using an incomplete query specification, for example by sketching the query.

One particular aspect of the sketch-based query is the difference between the modalities of sketches and 2D or 3D models. The core component of retrieval is matching, which in turn is dependent on the features being matched. Since differences in modalities will lead to differences in features, this means that multi-modal querying is especially challenging. Again there are extensive computer vision literature on multi-modal processing, examples of which included infra-red and visible images, range and visible images, aerial and satellite images that can potentially be helpful for suggesting approaches to sketch-based queries. When comparing multi-modal data, for example, for matching purposes, researchers follow four typical approaches as follows.

One approach is to extract a given type of feature vector from both modalities, and use a standard distance measure to match pairs of features. In general, this approach is unlikely to be successful since the different appearances of the modalities will result in different responses from the feature detectors and/or the feature descriptors, even if reasonably robust descriptors such as scale invariant feature transform (SIFT) [40] are used.

In some situations, it is possible to use the same features, but employ a specific distance measure that is able to cope with the difference in the modalities. For instance, in medical imaging the images from different sources (for example, magnetic resonance imaging (MRI) and computed tomography (CT)) need to be aligned, and many methods for multi-modal image registration has been developed, often based on mutual information [46].

A different approach is to map the images (or the features) from different modalities to a common space in which they are more highly correlated. This approach is commonly used for the recognition of artists' sketches for criminal investigations, in which a sketch of a suspect is matched to a mug-shot photograph. Many possibilities for constructing such mappings are available, for example, Partial Least Squares, a bilinear model [52], Canonical Correlation Analysis [36]. Following the mapping, matching should be relatively straightforward, and standard methods can be used.

A fourth possibility is to map one modality to the other. Again in the context of forensic sketch recognition, this would mean converting photographs in the gallery (database) to sketches or converting a sketch to a photograph [61]. This is the general approach that will be covered in this chapter: to bridge the gap between the modalities of sketches and 3D models, the latter will be converted to sketches, and in particular, images of sketches. In practice this means that some method for transforming 3D models will be required that can synthesize images that have the appearance of sketches.

There is a large body of work on transforming images in order to change their appearance, and in addition, there is a lesser amount of related work on transforming 3D models. Traditionally this work was termed non-photorealistic rendering (NPR) [35, 48] and operated at the intersection of computer vision and computer graphics. More recently, with the advent of deep learning, there has been an influx of activity, and a huge influx of papers, on neural image stylization [24], which shares some similar goals. While NPR's goals of creating artistic and expressive stylizations are not central to sketch retrieval, its goals of creating stylistic and abstracted stylizations are relevant. NPR has been applied to generate many different media types (for example, pen and ink, pencil, crayon, oil, watercolor), and for the proposes of sketch retrieval pen or pencil style renderings are likely to be most similar to a user's input query sketches. Abstraction is an important aspect of image stylization that can have a significant effect on how a stylized image is perceived or used. For example, Luft et al. [43] describe a system for creating renderings of rooms from CAD models, that by abstraction of shape and shading would provide more appealing visuals for customers. Another example is provided by Wu et al. [63] for an application of automatically generating bas-reliefs from single photographs of human faces. They found that incorporating NPR into the images as a pre-processing before generating the bas-relief improved the results. This was due to the suppression of unwanted detail as well as the emphasize of the important features. In terms of sketch retrieval the same principles apply: retrieval will be facilitated if confusing detail can be removed, while significant features must be retained and may even be emphasized.

There are two main strategies that can be followed in order to generate synthesized sketches from the 3D models. Either the sketches can be created directly from the 3D models, or else a 2D view-based approach can be employed. The former would generate 3D lines on the 3D surface, which can be rendered as multiple 2D views to enable them to be matched to 2D query sketches. The latter employs a traditional photorealistic rendering approach to render multiple views of the 3D model, to which image stylization methods are applied. In this chapter, we only consider contour and crease lines (described in more detail below), for which we can expect to find evidence directly from local features in the model or image. Gryaditskaya et al. [29] note that designers use these lines in their sketches, along with other inferred lines (cross-sections and various types of construction lines) during the design process. See also the curve networks extracted by FlowRep [27] illustrated below.

Figure 5.1 illustrates examples of how sketches drawn by a range of designers with different backgrounds appear for two 3D models. First, sketches from three designers are displayed in red, green, and blue, and shown overlaid in Figure 5.1(b) and Figure 5.1(f) [9]. Second, two further sketches

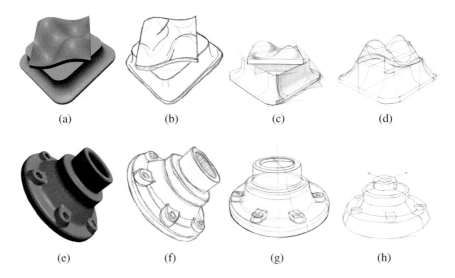

(a)	(b)	(c)	(d)
(e)	(f)	(g)	(h)

Figure 5.1 Two models and corresponding sketches; b) and f) overlay of three sketches from [9] displayed in red, green, and blue; c) and g) professional sketches from OpenSketch [29]; d) and h) student sketches from OpenSketch [29].

are provided from OpenSketch [29], drawn by a professional (Figure 5.1(c) and Figure 5.1(g)) and student (Figure 5.1(d) and Figure 5.1(h)). Whereas the first set of sketches was made using a reference image from the same viewpoint, for the second set of sketches the participants were only shown front, top and side views of the object, but required to create their drawings from a novel viewpoint.

Naturally, as can be seen, users have different drawing styles and that even within one drawing style there are variations in the precise placement of lines. In a complementary fashion, we note that there are also NPR algorithms that perform different line drawing stylizations, and that within a given stylization different algorithms will exhibit variations in detail. Subsequently, we note that there is not a single ideal set of ground truth sketches that can be used to evaluate the performance of alternative methods for line detection. The possibility of creating a different line drawing stylizations from any given model can be used to augment datasets of free-hand sketches required for training sketch-simplification and sketch-based modeling algorithms as discussed in Chapter 2 and Chapter 4 respectively.

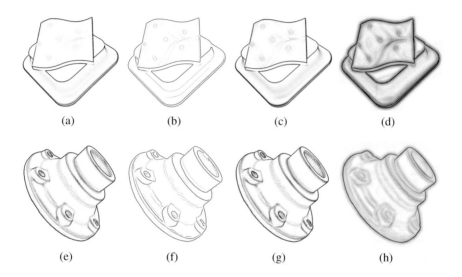

(a)   (b)   (c)   (d)

(e)   (f)   (g)   (h)

Figure 5.2   Detected image space edges; a) Sobel, b) Canny [6], c) morphological gradient, d) globalPb [2].

5.1 2D Image-Based Line Detection

In addition to NPR methods for extracting lines from images, many others have been developed in the broader area of computer vision for various applications. A related topic to line detection is edge detection, which we will also cover, since lines drawn by both artists and users sketching for retrieval could represent either lines or edges. In fact, Cole et al. [9] found that among the features they evaluated, image space intensity gradient magnitudes were the most effective feature for predicting the probability that an artist would draw a line.

5.1.1 Image-Based Edge Detection

A simple and classic method for extracting edges is the Sobel edge detector. It estimates horizontal and vertical gradients by convolving the image with two 3×3 kernels, and these are combined to a single overall gradient by taking the root of the sum of the squared directional gradients (see Figure 5.2(a) and Figure 5.2(e)). Some 15 years later this approach was refined by Canny [6]. Prior to calculating gradients, Gaussian blurring is first applied to the image, which enables larger scale edges to be detected. In addition, non-maximum suppression is used to thin the edges to a single, uniform thickness (see Figure 5.2(b) and Figure 5.2(f)).

The field of mathematical morphology provides a rather different approach to computing edges, known as morphological gradients [56]. The morphological gradient is defined as the difference between the dilation and the erosion of an image. Figure 5.2(c) and Figure 5.2(g) shows the morphological gradients obtained with a disk structuring element; the results appear similar to the Sobel edges.

The final method for edge detection that we will refer to is the global probability of boundary (globalPb) [2]. Unlike the above approaches, which are aimed at detecting edges from image intensities, globalPb aims to detect contours, and uses color and texture in addition to intensity. Briefly, local multiscale edge, color and texture features are extracted and combined using a supervised learning-based framework. In addition, they use spectral clustering applied to an affinity matrix describing similarity, between pixels to capture more global image information. However, for untextured CAD models such as those in Figure 5.1 there is little benefit to using the additional color and texture cues; see Figure 5.2(d) and Figure 5.2(h).

5.1.2 Image-Based Line Detection

We will describe several methods for extracting lines from images, covering traditional, NPR, and machine learning (including deep learning) approaches.

5.1.2.1 Traditional Approaches to Image-Based Line Detection

Many line detection methods have been developed in the broad area of computer vision for various applications such as medical imaging (for example, blood vessel extraction from retinal images) and remote sensing (road network extraction from aerial images). Like edge detection, some of these involve derivatives. For instance, Carsten [7] finds points with a zero intensity derivative and a high absolute second directional derivative perpendicular to the line direction. The direction of the line is determined from the eigenvalues and eigenvectors of the Hessian matrix, where the second-order derivatives are estimated from the Gaussian blurred image. Figure 5.3(a) and Figure 5.3(b) show the results of Carsten's method with additional pixel linking and hysteresis thresholding. The bright lines corresponding to ridges are plotted in red, while the dark lines corresponding to valleys are plotted in blue. The hysteresis thresholding approach is popular in edge detection and can be applied to improve line detection: two-line

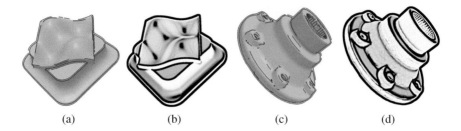

(a)   (b)   (c)   (d)

Figure 5.3 Detected image space lines; a) and c) 2D ridges (red) and valleys (blue), b) and d) Frangi et al. [18].

response thresholds are applied, and those pixels above the high threshold are retained as lines, while those pixels below the low threshold are discarded. Pixels with intermediate line responses between the thresholds are only retained if they are connected to pixels above the high threshold, which were therefore determined to form NPR, lines. Another approach based on the Hessian, designed to detect blood vessels, is given by Frangi et al. [18]. A measure is constructed from the eigenvalues of the Hessian, and is computed at multiple scales. The maximum response over all scales is taken as the final vesselness measure (Figure 5.3(c) and Figure 5.3(d)).

5.1.2.2 Machine Learning Approaches to Image-Based Line Detection

Rather than manually designing line detectors, an alternative approach is to learn a filter based on training examples. Soares et al. [55] tackle retinal vessel segmentation (that is, two classes: vessel or non-vessel) using a bank of Gabor wavelets provide features covering different scales, frequencies, and orientations. These are then fed into a Bayesian classifier. The data consisted of retinal images with corresponding manually derived binary segmentations. Figure 5.4 gives results using a similar approach, except that the responses from the Gabor bank are combined using generalised linear model (GLM) regression. Figure 5.4(a) and Figure 5.4(e) show the direct output from a model that was built from the 20 training images provided in the digital retinal images for vessel extraction (DRIVE) dataset. In addition, a receiver operating characteristic (ROC) curve was constructed from the training data, and the optimal threshold that maximized Youden's index was determined, and applied to the test images (see Figure 5.4(b) and Figure 5.4(f)).

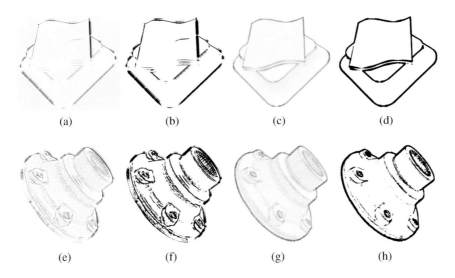

(a)   (b)   (c)   (d)

(e)   (f)   (g)   (h)

Figure 5.4 Detected image space lines (before and after thresholding); GLM model trained on data from DRIVE (a,b, e, f) and Cole et al. [9] (c, d, g, h).

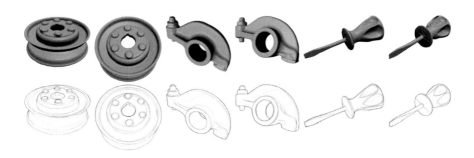

Figure 5.5 Examples of training data pairs constructed from the drawings in Cole et al. [9].

With such a machine learning approach it is straightforward to retrain if alternative data is available. As an example, we use the dataset from [9], which they had carefully constructed in order to provide a comprehensive set of objects with specific characteristics including objects that contained certain geometric features of interest, were reasonably simple, and were familiar to artists among others. The objects consisted of bones, tablecloths, mechanical parts, and synthetic shapes. For each of the 3D models, two viewpoints and

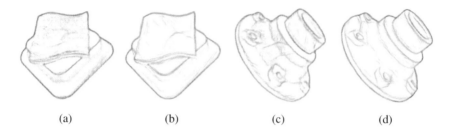

(a)   (b)   (c)   (d)

Figure 5.6   Image space lines generated by pix2pix [30]; a) and c) initial results, b) and d) results using additional data augmentation.

two lighting conditions were provided, and drawings were collected from 29 artists (student and professional). To train the GLM model, the two 3D models shown in this chapter's figures were excluded and for each of the two views of the remaining 10 models the drawings were averaged, forming 20 training images paired with the rendered 2D views. Whereas the ground truth for the retinal images consisted of binary images, this new ground truth is not binary; some examples are shown in Figure 5.5. This means that an appropriate threshold could not be learned from the training data as was done for the DRIVE dataset. Standard image thresholding algorithms gave poor results on Figure 5.4(c) and Figure 5.4(g); the current reasonable results (Figure 5.4(d) and Figure 5.4(h)) were achieved using thresholding [62] of the foreground region followed by hysteresis to recover from the initial over-thresholding. The results of applying the retrained regression model (Figure 5.4(c) and Figure 5.4(g), and Figure 5.4(d) and Figure 5.4(h), before and after thresholding respectively) are similar to the results from the DRIVE model, indicating that the line detection methodologies which have been developed for other applications (medical image analysis, remote sensing, etc.) are reasonably appropriate for extracting lines from views of 3D models. The result for the latter GLM model for the first object (Figure 5.4(d)) is significantly better than the former GLM model (Figure 5.4(b)). For the second object, the latter GLM model produces a much cleaner result consisting of fewer spurious lines, although some object structures have also been lost. Cole et al. [9] also took a machine learning approach to predict where people draw lines on 3D objects and built a regression model on various features, both 2D image-based and 3D model-based.

5.1.2.3 Deep Learning Approaches to Image-Based Line Detection

Figure 5.6 shows an alternative machine learning approach, using deep learning, and specifically pix2pix [30], which has shown great success in many diverse applications of style transfer (so called image-to-image translation). pix2pix uses a conditional generative adversarial network (CGAN) along with a "U-Net" architecture (an encoder-decoder with skip connections between corresponding layers in the encoder and decoder stacks) and a PatchGAN classifier for the discriminator. One limitation of deep learning is that large amounts of training are generally required to achieve good results. The initial results of training pix2pix using the same 20 training images described above are quite noisy, with both missing structures and spurious lines, see Figure 5.6(a) and Figure 5.6(c). In order to improve these results, the training set was enlarged to 160 images using data augmentation (in addition to the built-in pix2pix data augmentation of rescaling and random cropping), specifically by creating versions of the images at multiples of 45°. This is valid as we expect the line detection to be orientation invariant. As seen in Figure 5.6(b) and Figure 5.6(d) this provides better quality lines, which could be further improved if more training data was available.

Further examples of deep learning approaches to line detection are given in the next section.

5.1.2.4 NPR Approaches to Image-Based Line Detection

For NPR, the emphasis is more on creating coherent lines rather than focusing on the accuracy of line detection. In other words, the lines should be smooth extended curves, and not consist of fragmented groups of pixels. Kang et al. [32] achieved this using a difference of Gaussians (DoG) filter. While the DoG (as an efficient approximation to the Laplacian kernel) is often used for blob or line detection, their improvement was to replace the standard fixed square convolution kernel with an adaptive kernel. The first estimate the local image direction (the edge tangent flow) by taking the 2D vector field of vectors perpendicular to the intensity gradients, and apply smoothing to reduce the effect of noise, which is particularly present in areas with low gradient magnitude. The smooth direction field is used to steer the DoG filter by deforming the kernel to align with the salient image features. As can be seen in Figure 5.7(a) and Figure 5.7(e) this tends to produce a coherent set of lines.

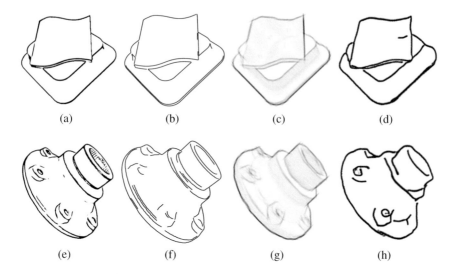

(a)   (b)   (c)   (d)

(e)   (f)   (g)   (h)

Figure 5.7   Detected image space lines; Kang et al. [32], von Gioi and Randall [59], Lu et al. [41], Li et al. [39]

Although not designed for NPR, the work by von Gioi and Randall [59] also aims to avoid fragmented and irregular curves, even at the cost of missing some structures in the image, and so we include it in this section. They follow the 'a contrario framework' such that non-accidental arrangements of parts of curves should be retained. They use the Canny edge detector to generate candidate curves, and then the Mann-Whitney U statistical test is applied to check if the intensities in the neighborhoods on either side of the curve are significantly different. The statistical test is applied to parts of those curves that can be locally approximated by circular arcs, and the smoothness arises since the circular arcs approximating a contour must be large enough to satisfy the statistical test. Indeed, the results in Figure 5.7(b) and Figure 5.7(f) are fairly clean, although some structures are missing.

It was mentioned that there are variations in the placement of lines between users. While this is in part due to different interpretations of objects, it is also indicative that users do not always carefully delineate features. Some NPR techniques also consider and emulate this imprecise and sketchy nature of drawing. For instance, Lu et al. [41] describe an algorithm for generating pencil drawings and creating slightly lengthened lines to provide a sketchy impression (see Figure 5.7(c) and Figure 5.7(g)). Some parts of their pipeline are not so relevant to sketch-based queries: the first stage is to detect lines whilst trying to avoid false responses due to clutter and texture in the image. Lu et al. first perform convolution using as kernels a set of eight-line segments in the horizontal, vertical, and diagonal directions. This produces eight response maps, from which each pixel is classified into the direction that gives the maximum response, and the remaining response maps are zeroed at that position. The second set of convolutions is applied to the modified eight response maps with the eight-line kernels. The effect of these two sets of convolutions is to suppress responses from noise, and also to draw lines that link disconnected gradients. While noise and clutter are not likely to be primary issues for sketch-based queries, bridging gaps in gradients will produce a more coherent set of lines, which is desirable. The final part is to generate a tone map (that is based on a remapping of the intensities in the input image), and this is combined using a simple product of the line and tone maps. Again, the tone map is of less concern to sketch-based queries, although as we saw in Figure 5.1 some users may like to include tonal values in their sketch. In our example in Figure 5.7(c) and Figure 5.7(g) the contrast of the tonal element has been reduced in order to emphasize the lines. It is also possible to take a deep learning approach to pencil stylization of images, and Gao et al. [23] describe an approach in which Lu et al.'s pipeline is used to generate training data for a CNN. When creating this training data they manually adjust several of the parameters such as the length of the line segment kernels, for example, to enable the traditional method to better cope with variations in the images. Consequently, the resulting CNN is more robust than Lu et al.'s original method.

We show a final image-based method that allows for, and encourages, a sketchy effect. Li et al. [39] also take a deep learning approach, using a generative adversarial network (GAN) with a standard encoder-decoder architecture. The loss function is modified to cope with multiple training targets that are, for example, drawings based on an image provided by multiple users, while a global GAN is used for the discriminator rather than PatchGAN, which they report helps the network preserve continuous lines. Li et al. claim that this enables their model, like a non-professional artist, to generate lines that are not perfectly straight. As the results in Figure 5.7(d) and Figure 5.7(h) demonstrate, although some of the object's structure is not captured, the lines do have a realistically sketchy appearance including the distorted lines.

If we look at all the above results we can see that the second 3D model (the flange) is more problematic than the first, since the smooth curve along the middle of the collar is not picked up well by most of the methods. Not

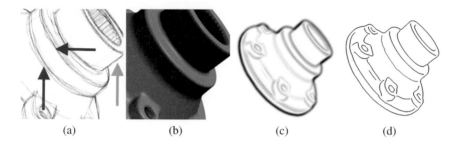

      (a)   (b)   (c)   (d)

Figure 5.8   Capturing indistinct image space lines; (a) averaged multiple user drawings, (b) rendering artifacts, (c) Canny edge response at $\sigma = 8$ without non-maximal suppression, (d) extracted Canny edges.

surprisingly the five user sketches (a zoom-in of their average is shown in Figure 5.8(a)) are not as consistent for this curve (indicated by the red arrow) as for some of the other curves such as occluding boundaries (green arrow). However, this is not the main source of difficulty, as other user-drawn lines such as the ridge (blue arrow) parallel to the problem line (that is, the curve along the middle of the collar) are equally dispersed but are drawn well by most methods. Looking closely at the rendered view of the model (Figure 5.8(b)) some minor artifacts are visible in highly curved sections of the surfaces. Although this is present in the areas pointed to by both the red and blue arrows, the higher contrast at the ridge (blue arrow) enables that line to be drawn well by most methods. Thus, although we previously noted that there are variations amongst user drawings and also variations in the stylizations of different algorithms, we should also consider variations in the appearance of the 3D model due to different shaders. However, it is possible to tune the detectors to try to capture these problematic features. For example, rerunning the Canny edge detector at a larger scale ($\sigma = 8$ instead of $\sigma = 4$ which was used for the result in Figure 5.2) enables the fragmented responses produced at $\sigma = 4$ to be merged into a single band response that is continuous and coherent, although wide and diffuse, as well as a rather lower magnitude than those for the other lines (Figure 5.8(c)). Nevertheless, after non-maximal suppression and hysteresis thresholding are applied, the curve is successfully extracted (highlighted in red) – see Figure 5.8(d).

5.1.2.5 Issues with Image-Based Line Detection Methods
The above illustrates the challenge of detecting subtle lines. Another issue is the quality of the extracted lines. It was noted that one goal of NPR is

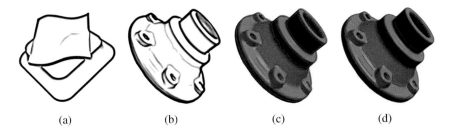

(a)   (b)   (c)   (d)

Figure 5.9 Issues of positional accuracy; (a,b) multiscale-fractional anisotropic tensor (MFAT), (c) MFAT for the second model thinned and overlaid on the rendered image, (d) overlay of Canny edges from Figure 5.8.

to produce coherent lines, but the outputs of many lines and edge detectors are often noisy and fragmented. Therefore many postprocessing methods (for examples, sequential search [16], mathematical morphology [19], hysteresis thresholding [6, 7], deep learning [49]) have been developed to improve their quality, by filling gaps and removing isolated false positives. Yet another consideration is localization error. For instance, Figure 5.9(a) and Figure 5.9(b) shows results from Alhasson et al. [1] vessel detection method (namely, the MFAT). Like Frangi et al. [18] and others, it is based on eigenvalues of the Hessian at multiple scales and has some modifications to cope better with background noise and junctions. The results are quite clean (coherent) and most of the important lines are captured. But when the extracted lines for the second 3D model are overlaid on the rendered view in Figure 5.9(c) (they have been thinned to make it easier to check their position) it shows that there is some mislocalization; lines delineating the outer edges are shifted inwards, while the tricky collar line in the flange discussed above is shifted upwards into the dark region. In comparison, the Canny edges overlaid in Figure 5.9(d) are more accurate for this task. It is worth noting that users' drawings may also exhibit mislocalization; Schmidt et al. [50] found that even expert 3D artists are prone to perceptual biases and that the 3D error in their drawings depends on their viewpoint with respect to the object's surface.

5.2 3D Model-Based Line Detection

As was shown above, many of the existing image-based line detection methods are fairly effective and robust. However, by operating on images they have to cope with incomplete and confusing information, whereas if lines are extracted from 3D models rather than synthesized views of these models, then these lines are more likely to directly reflect the true geometry of the objects. This potentially makes 3D model-based approaches to line detection more reliable since 2D image-based line detection can be affected by extraneous factors, such as illumination, textures, and perspective distortion, meaning that significant lines may easily be missed, and spurious lines introduced. Bénard and Hertzmann [5] provide a thorough description of methods for generating line drawings from 3D models, covering details of the underpinning theory as well as computations for different 3D representations such as meshes, parametric surfaces, and implicit surfaces. In this section we provide a brief overview of several methods for locating lines on the surface of a 3D model.

5.2.1 Detection of Separate Lines

A straightforward approach is to find locations with extremal principal curvature in the principal direction – such loci are often called ridges and valleys (and sometimes crests or creases) – and often indicate salient structures of the object. The curvature of a surface is an intrinsic property, and thus, the ridge and valley lines exist in the same location of the surface, independent of the view. Moreover, curvature and the majority of 3D model-based line detection methods are independent of the lighting conditions. Figure 5.10(a) and Figure 5.10(e) shows the ridges and valleys extracted from the 3D models, which are drawn as red and blue lines respectively.

Many definitions for contours are view-dependent, that is, their position on the 3D surface depends on the viewpoint, and it has been argued that view-dependent lines better convey smooth surfaces [13]. Another straightforward definition for lines, which unlike ridges and valleys is view-dependent, is for occluding contours. These are the projection onto the image plane of the rim, that is, points on the surface whose normal is perpendicular to the viewing direction. In other words, for a given viewpoint the occluding contours will be the depth discontinuities that delineate the separation between the object and background as well as delineating self occlusions. Such contours are known to be important for human visual perception [58]. However, for many models such as those shown in the above examples, there will be few occluding contours beyond the outer silhouette.

A related concept is suggestive contours [13], which are locations at which the surface is almost in contour from the original viewpoint, and can

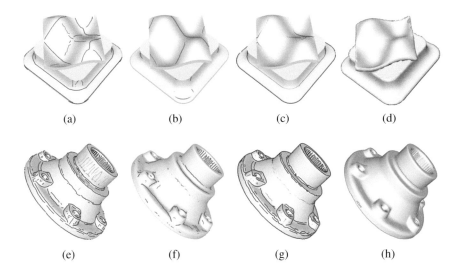

Figure 5.10   Detected 3D space lines; 3D ridges (red) and valleys (blue), suggestive contours, apparent ridges, relief edges.

be considered to be locations of true contours in nearby viewpoints. Thus it is a relaxation of the strict definition for occluding contours. The definition for suggestive contours is the set of locations at which the dot product of the unit surface is normal and the view vector is a positive local minimum rather than zero. Examples are shown in Figure 5.10(b) and Figure 5.10(f).

Judd et al. [31] proposed a view-dependent version of ridges, named apparent ridges. They defined a view-dependent measure of curvature based on how much the surface bends from the viewpoint. Thus, it takes into consideration both the curvature of the object and the fore-shortening due to surface orientation. Apparent ridges are then defined as locations with maximal view-dependent curvature in the principal view-dependent curvature direction. As can be seen in Figure 5.10(g), while many of the ridgelines in Figure 5.10(e) are retained, some are removed such as those on the outer surface of the cylinder at the top, as well as the curve along the middle of the collar (highlighted in red in Figure 5.8(d)).

Whereas the above approaches have all aimed to extract curves from 3D models that capture their overall structure, a different approach is to focus on local geometric details, such as decorations, that could be distinctive. In the context of archaeological artifacts, Romanengo et al. [47] extract such curves

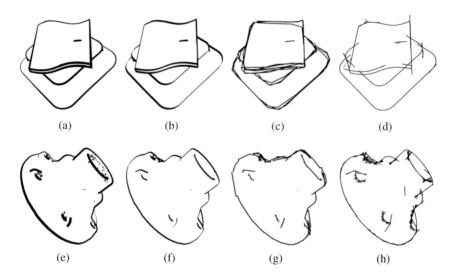

(a)   (b)   (c)   (d)

(e)   (f)   (g)   (h)

Figure 5.11   Some 3D rendering styles produced by Blender; a) and e) distance from the camera, b) and f) calligraphy style, c) and g) polygonization style, d) and h) backbone style.

by identifying characteristic points which are approximated by curves. Local properties are thresholded to extract significant feature points (that is, with high and low values); mean curvatures values are used for geometric features, and luminosity is used for color decorations.

Another approach to extracting relief edges was developed by Kolomenkin et al. [34] for the visualization of archaeological artifacts. They consider surfaces to be made up of a relatively smooth base surface with a superimposed relief, and they treat this local height function as an offset image. Relief edges are then the edges of this image, and are defined as the zero crossings of the normal curvature in the direction perpendicular to the edge, and are therefore view-independent. A local polynomial model is fitted to the data to estimate the edge direction and magnitude, and a rough estimate of the normal direction of the base surface is used to help constrain the possible orientations. Since the two 3D models that we are using have almost no surface detail beyond the grooves inside the cylinder in the second model, it is not surprising that there are few relief edges detected (Figure 5.10(d) and Figure 5.10(h)).

In addition to the above, there are many other approaches to generating stylizations of lines from 3D models. A general framework was described by Grabli et al. [28]. They incorporate control over the style attributes of strokes (topology, geometry, thickness, color, etc.) and also use an estimate of density to omit lines (to avoid clutter or focus attention). Some 3D computer graphics packages include NPR rendering engines, which enable some of these stylizations. We show some examples from Blender's Freestyle rendering mode in Figure 5.11. It uses Z-depth buffer information to draw lines on selected edge types (for example, occluding boundaries, crease edges whose dihedral angle is less than a threshold, as well as ridges and valleys and suggestive contours). Two styles in which the thickness is modified are either according to the distance from the camera or according to the orientation of the stroke (calligraphy style), and are shown in Figure 5.11(a) and Figure 5.11(e), and Figure 5.11(b) and Figure 5.11(f) respectively. A more sophisticated version (not incorporated into Blender), the so-called "isophote distance" also modifies thickness depending on depth, curvature, and light direction [26]. The second pair of styles are examples of a more sketchy approach. Polygonization (Figure 5.11(c) and Figure 5.11(g)) simplifies strokes, and using multiple rounds (three in this example) creates the sketchy style by generating multiple strokes with random perturbations. The Backbone Stretcher style (Figure 5.11(d) and Figure 5.11(h)) extends the beginning and end of each stroke, reminiscent of the extended pencil drawing lines of [41] that were shown in Figure 5.7.

In contrast to image-based line detection, there has been substantially less take-up of deep learning for 3D models due to the irregular structure of 3D data. Thus, a common approach is to process 3D as multiview 2D data. For instance, Ye et al. [64] take this approach. They aim to generate line drawings from 3D objects but apply their model to 2D images rendered from the 3D object. Their GAN incorporates long short-term memory (LSTM) to enable it to generate lines as sequences of 2D coordinates. Nguyen-Phuoc et al. [45] propose a CNN-based differentiable rendering system for 3D voxel grids which is able to learn various kinds of shader including contour shading. While voxel grids provide the regular structure missing from meshes, in practice their large memory requirements limit their resolution, and so they resize their 3D data to a $64 \times 64 \times 64$ voxel grid.

5.2.2 Curve Networks

In the area of geometric modeling, the concept of a curve network is used to provide a descriptive and compact representation of a 3D model by extracting a network of 3D curves that lie on the model's surface, and are typically aligned with principal curvature directions. They have found use as tools for shape editing, simplification, etc. Methods for extracting curve networks are influenced by concepts from perception and design literature, as well as computer graphics, and 2D projections of the network should be indicative of the shape's 3D geometry. Therefore such curve networks should be well suited to providing appropriate stylizations of shapes for sketch retrieval.

One such method is the Exoskeleton [12], which performs a coarse abstraction of 3D shapes that aims to capture the important structures, see Figure 5.12. Unlike the above methods, it uses a segmentation of the 3D model into roughly convex parts to provide information about the structural parts; in particular, they consider the segmentation boundaries as perceptual features. Their examples include results from both manual and automatically generated segmentations. From the initial segmentation they use normal-based features to subdivide these regions into patches, whose boundaries follow perceptual and normal-based lines. The Exoskeleton algorithm provides some flexibility in how the segmented regions are processed so that different aspects of shape can be emphasized, for example, the symmetry axes. In order to generate pleasing results, they attempt to generate curves that exhibit the following desirable properties:

1. the normal-based lines follow the principal curves,
2. they are well spaced, that is, their density should be proportional to the local complexity of the shape,
3. the curves should be long and continuous, and
4. the normal-based lines should be as orthogonal as possible to the perceptual lines.

Their results look promising, at least for simple clean 3D models, although feature curves are inaccurate within transition regions with weak features [42].

Gehre et al. [25] describe a different approach to extracting curves from surfaces that also consider merging or suppressing nearby geometric feature curves, but in this case depending on a global scale parameter.

FlowRep, proposed by Gori et al. [27], is another method for extracting curve networks. They first extract sequences of the input mesh edges that form long, well-shaped flowlines. Next, there is a sequence of steps that refine the inadequately described cycles in the initial network by first adding to the network those flowlines that cross these cycles, followed by then removing redundant flowlines. The resulting curve networks are explicitly intended to be projectable, that is, their 2D projections indicate the shape's

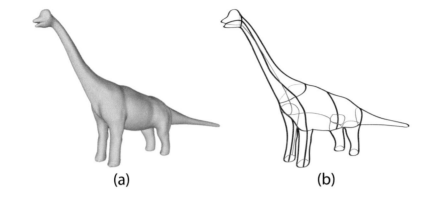

(a)   (b)

Figure 5.12 A curve network extracted using Exoskeleton [12].

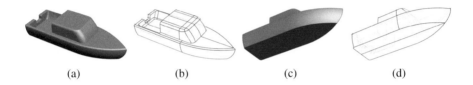

(a)   (b)   (c)   (d)

Figure 5.13 A curve network extracted using FlowRep [27]

3D geometry. This is achieved by incorporating a projectivity term that is included in the overall cost that is optimized. The projectivity of a curve is computed according to its planarity, its deviation from local geodesics (which is evaluated using the angle between the normal to the surface and a local plane fitted to the curve), and a term to encourage orthogonal crossings within the network. One may observe that the concept of using projectivity, planarity, and orthogonality as an optimization cost for the projection of the 3D object onto 2D lines is similar, in principle, to the adoption of regularities in the reconstruction of 2D sketches as discussed in Section 4.2. An example showing two views of a curve network extracted using FlowRep are given in Figure 5.13.

5.3 2D Image-Based or 3D Model-Based Line Detection?

It was stated above that Cole et al. [9] built a regression model using both 2D image-based and 3D model-based features. Also, for different types of objects (bone, cloth, mechanical and synthetic) they constructed random

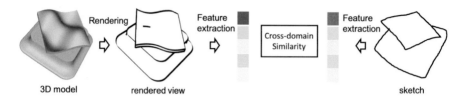

Figure 5.14   Pipeline of a sketch-based retrieval system.

forests for predicting the positions of lines and used the relative importance of these features, which was provided by the forests, to analyze the results. Surprisingly, they found that image space features, in particular intensity gradient magnitude, were more effective than those directly extracted from the 3D geometry (although the latter did nevertheless contribute some useful information). Moreover, they showed that for different types of objects, different features were more predictive of their artist drawn lines. For instance, the mechanical models lines were best predicted using ridges and edges, while the cloth and bone model lines were best predicted using occluding contours and suggestive contours. Thus, there is, as yet, no clear winner between the 2D image-based or 3D model-based approaches to line detection. Moreover, many of the 3D model-based approaches use view-dependent quantities rather than rely purely on the intrinsic geometry of the 3D object.

5.4 Sketch-Based 3D Shape Retrieval

We have studied how 3D shapes can be rendered as sketched drawings, and will now describe how these rendered views can be used to facilitate searching for these shapes in a database. This is based on the insight that if two shapes look similar then they are likely to be similar.

Sketch-based shape retrieval has been widely studied, especially after the introduction of the Princeton Shape Benchmark [53], where a system is proposed to take as input the side view, front view, and top view of a 3D object. Then the algorithm retrieves 3D objects that closely agree with the given views. There exists significant prior research such as Chen et al. [8], Yoon et al. [65].

The main difficulty of sketch-based shape retrieval lies in the difference between two feature domains, that is, the feature space of 2D sketches and 3D shapes. Traditional methods usually consist of the following pipeline:

view selection and rendering, feature extraction [4], metric learning, and matching as shown in Figure 5.14. Efforts were made for more effective descriptors of both sketches and shapes. To facilitate the feature extraction, 3D shapes are usually represented as a set of rendered views, which is called the "multi-views" method. Then image-based features can be applied to these views. Traditional methods try to minimize the difference between the two domains by rendering 3D shapes in hand-sketch styles using the aforementioned NPR techniques. With the development of deep learning, more and more methods have been proposed for multi-modal applications. These methods usually map data of different modalities into one common feature space where cross-domain similarity is calculated. In our problem, the 2D sketches and 3D shapes are collectively mapped into a common feature space where the retrieval is conducted. As to variations in orientation between the query and target, traditional methods typically use rotation-invariant features to cope with this, while deep learning methods employ data augmentation (generating multiple instances by rotating the original samples) to handle such variations. We will introduce both types of methods in the following.

5.4.1 Traditional Approaches

Early research usually focuses on hand-crafted features. It always consists of a feature extractor and a similarity measurement. In the following we will introduce several representative methods of traditional features.

5.4.1.1 Fourier Descriptors

Fourier descriptors are commonly-used descriptors specifically designed for closed curves [66]. By decomposing curves into different frequencies, the descriptor is both efficient and well studied. However, it normally only considers the outer boundary, and the interior is ignored which involves a loss of information.

5.4.1.2 Shape Context

Shape context [4] calculates a histogram of all surrounding samples around a sample point. It presents these points in a log-polar space and the space is partitioned into 5×12 individual bins with 5 different distances and 12 different orientations. Each sample point falls into one of those bins as shown in Figure 5.15. It is empirically demonstrated to be scale and transformation-invariant.

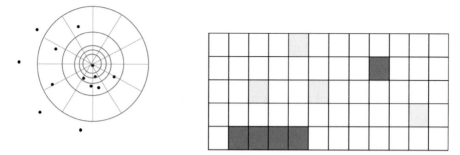

Figure 5.15   Shape context descriptor. Left: scattered points. Right: feature calculated.

5.4.1.3 Lightfield Descriptors

The lightfield descriptor [8] is one of the most commonly-used hand-crafted features used for shape retrieval. It first takes pictures of a given 3D model by putting cameras on 20 vertices of a regular dodecahedron. These pictures are used to roughly represent the model. Zernike moments and Fourier descriptors are then employed for feature extraction for these rendered silhouettes. A brute force search is then applied to match the corresponding views. The similarity is calculated as the sum of similarities of all corresponding views. The L_1 distance[1] is used to compute image feature distance.

To make the rotation estimation more accurate and robust, a set of lightfield descriptors are extracted instead of a single one as shown in Figure 5.16. The dissimilarity between two models is defined as the minimum distance between their corresponding lightfield descriptors.

5.4.1.4 Boundary-Based Descriptors

A 3D search engine was proposed in 2003 [20]. This engine supports both 2D sketch and 3D model queries. To facilitate sketch-based retrieval, the method used 2D spherical harmonics [33] as features. This method first converts a boundary contour into a distance function and circular functions are generated based on different radii. Features are then extracted as amplitudes on different trigonometric functions.

---

[1] Also known as Manhattan Distance. L_1 Norm is the sum of the magnitudes of the vectors in a space.

Figure 5.16 A set of Lightfield descriptors for a single model Chen et al. [8].

5.4.1.5 Diffusion Tensor Fields

Diffusion tensor fields were originally introduced to measure the anisotropic properties of water diffusion in MRI medical imaging [3]. They can be applied to sketch images to estimate the orientation of each pixel depending on its surrounding pixels [65]. Features are represented as histograms of magnitude and orientation of the ellipsoidal representation of each contour pixel. Similarity of two histograms H_i, H_j of diffusion tensor fields can be obtained by S:

$$S(H_i, H_j) = \frac{H_i \cdot H_j}{\|H_i\| \|H_j\|} \qquad (5.1)$$

5.4.1.6 GALIF Descriptors

The Gabor local line-based feature (GALIF) [15] was proposed in 2012 as part of the design of a sketch-based retrieval system. GALIF defines a filter bank f of Gabor functions (see Box 2.2 for a description of the Gabor filter) with different orientations as the following expression in the frequency domain:

$$f(u, v) = exp(-2\pi^2((u_\Theta - \omega_0)^2 \sigma_x^2 + v_\Theta^2 \sigma_y^2)), \qquad (5.2)$$

where (u_Θ, v_Θ) are the rotated bases in the frequency domain. ω_0 is the peak response frequency. σ_x and σ_y are frequency and angular bandwidth respectively; see Figure 5.17.

These filters are then applied to the sketch, forming different response images for each individual orientation. Keypoints are uniformly sampled from the response images. For each keypoint, GALIF considers its

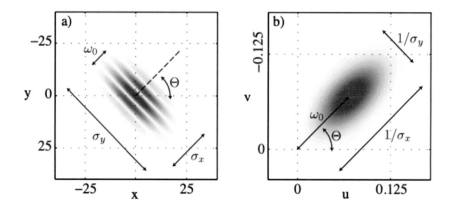

Figure 5.17   Gabor filters. Image from Eitz et al. [15].

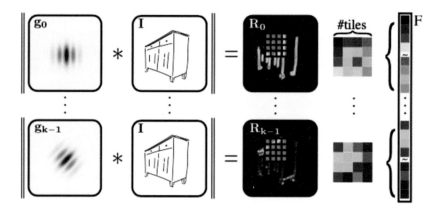

Figure 5.18   GALIF features. Image from Eitz et al. [15].

surrounding $n \times n$ neighbor cells and obtains a feature vector by averaging and concatenating the features of the cells (Figure 5.18).

Since GALIF is a local descriptor, Bag-of-features [54] is used to integrate them into a global feature.

5.4.1.7 Histogram of Oriented Gradients

Histogram of oriented gradients (HOG) was first proposed as a method for detection of humans [11]. It calculates the histogram of the gradients within each cell of an image in terms of both orientation and magnitude. To use HOG in sketch retrieval, Eitz et al. [14] proposed a simplified version of HOG only considers orientation.

5.4.1.8 Point-Based Registration

Some of the methods treat sketch and contour images as a set of sample points [51, 37]. The similarity is calculated as the distance between each aligned individual point. For instance, Shao et al. [51] first vectorize a user-provided sketch as polylines. Then samples are drawn within these polylines forming a set of points. The similarity between a sketch S and a contour T is computed as the sum of all distances between $p \in S$ and polylines in T.

5.4.1.9 Cross-Domain Manifold Approach

To tackle the multi-modality mapping problem, many cross-domain manifold approaches were proposed which try to establish correspondence between two different domains. For instance, Furuya and Ohbuchi [22] established two separate manifolds for sketches and 3D models. For each manifold, they employed different similarity metrics, that is, the GALIF [15] feature measurement is used for both sketch-model and sketch-sketch domains. The dense scale invariant feature transform (DSIFT) [21] feature is used for inter-model measurement. As mentioned before, GALIF is a bag-of-features based local descriptor. DSIFT is an appearance-based similarity measurement which first renders depth images for a 3D model, from which SIFT [40] features are extracted. Having similarities, the method is able to build a cross-domain manifold graph with weighted edges. Similarities between a query sketch and the 3D models can be calculated by employing a manifold ranking algorithm [67]. The insight of using a cross-domain manifold ranking is that similarity computed on the low-dimensional manifold is expected to be more accurate than in the original space.

5.4.2 Deep Learning Approaches

With the huge success of deep learning in many computer vision tasks, more and more researchers have started to look at the possibility of extending it to 3D data. As commonly known, deep neural networks produce superior features to hand-crafted features. Therefore many attempts have been made to achieve better results for sketch-based retrieval by employing deep networks. The main focus of these methods lies in how to effectively process the multi-modality data. In this section, we will introduce some of the representative works of applying deep networks to sketch-based 3D model retrieval.

5.4.2.1 Multi-View CNNs

It is a commonly used method to represent 3D shapes as a set of views in shape retrieval tasks [38]. This is also named the multi-view method. The first method tries to aggregate multiple views of a shape to form a compact feature using convolutional networks, and is known as Multi-view CNNs [57]. After extracting deep features for each individual view, the Multi-view CNNs aggregate these views by an operation called "view pooling", which is not much more than a max-pooling layer. For the purpose of sketch-based retrieval, the method produces Canny edges on the depth maps from 12 viewpoints. The aggregated features of these edge maps are used for the subsequent retrieval.

5.4.2.2 Siamese Networks

Wang et al. [60] proposed the first convolutional neural network for sketch-based 3D model retrieval method. The idea is to jointly learn a cross-domain metric for both views and sketches. Similar to most of the previous approaches, it first renders 3D models from different views. 2D line drawings are then achieved by employing closed boundaries and suggestive contours. Having the line drawing views v_i and sketches s_i as training samples, the method trains two Siamese networks, one of which extracts features for the sketch domain and the other for view features. Given two pairs of training samples $(s_1, s_2), (v_1, v_2)$, the loss function consists of three separate losses: loss of (s_1, s_2), loss of (v_1, v_2) and the cross-domain loss (v_1, s_1). If two samples are from the same class, the loss is proportional to their feature distance. Otherwise, the loss increases as the feature distance decreases. This operation is aimed to map samples from the same class as close as possible while pushing samples from different classes away from each other in the feature mapping process. The joint learning of two domains facilitates the cross-domain metric, by using which views and sketches can be mapped into a common feature space.

5.4.2.3 Cross-Domain Networks

Similar to the idea of cross-domain manifold methods in traditional approaches, deep learning methods also try to learn a joint feature space that

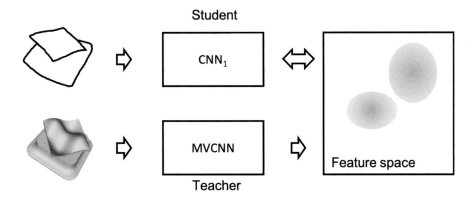

Figure 5.19   Cross-domain guidance networks.

suits both modalities. Dai and Liang [10] proposed a teacher-student network based on the idea of knowledge distillation (Figure 5.19). The feature space of 3D shapes is learned by the teacher network. The student network extracts features of 2D sketches with the guidance of the pre-computed feature space of 3D shapes. For the teacher network, 3D shapes are first represented as multi-views. Those rendered views are then used to train the CNNs. Average pooling is used to fuse views, forming a feature vector for each shape. The center of each class is calculated as the mean vector of all the shape features belonging to the same semantic category. When training the student network, the computed feature of a sketch is compared against its corresponding class center in shape feature space. This distance is set as the loss function which the network tries to minimize.

Different from methods of establishing common feature spaces, Mingjia et al. [44] tries to directly obtain a cross-domain mapping by leveraging Cycle CNNs without an explicit common feature space. In the training stage, the proposed method takes as inputs sketch-shape pairs. It consists of two CNNs for sketch and 3D shape respectively and two other CNNs that map sketches to shape features and the back again. Similar to Dai and Liang [10], 3D shapes are also represented by multi-views for the shape encoder. The loss function consists of three terms: cycle consistency loss, classification loss, and similarity loss. The cycle-consistency ensures the identity transform when the mapping and reversed mapping are both applied. The classification term enforces the mapped feature to belong to the corresponding category.

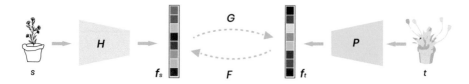

Figure 5.20   Cycle CNNs Mingjia et al. [44].

The distance between a mapped sketch feature and the corresponding shape feature is minimized by the similarity term.

5.5 Conclusion

It was noted that one of the challenges of applying sketch-based queries to retrieve 2D or 3D models is that the style of sketches is typically not the same as views of models. This effectively leads to the processing of multi-modal data, which is a task extensively explored in computer vision. One solution, covered in this chapter, consists of transforming the database of models in order to make them look more sketch-like, thereby facilitating the matching between query and database.

Both image space and object space (that is, 2D image-based and 3D model-based) approaches to the detection of lines have been described and illustrated with examples. The majority operate by extracting lines that relate to directly visible lines, such as contour and crease lines. However, some methods (for example, for curve networks) also generate other curves such as construction lines, which have to be inferred.

Although there is a wide range of methods for line extraction, there remain challenges in obtaining high-quality, accurate, continuous lines. Some lines can be relatively easily detected (for example, occluding contours), whereas others are more subtle. Figures 5.8 and 5.9 show examples of these issues. We can expect to see further development in this area, especially deploying deep learning.

Methods of retrieving 3D models by sketches have also been described. To achieve this, 3D models are usually rendered as multi-view images by the described line detection approaches. They are then compared with the provided sketch for retrieval. The main difficulty is the cross-domain feature extraction and matching. We roughly divided them into traditional methods which use hand-crafted features and deep learning-based ones which achieve a common feature space by leveraging specifically designed neural network structures.

The performance of sketch-based retrieval is a result of both line detection and the subsequent searching scheme. Superior line detection approaches naturally make the rendered views more similar to the sketch style, and hence make the subsequent searching stage easier. A strong matching algorithm should, however, bridge the gap by establishing a common feature space shared by the two modalities. Although there has been much effort in trying to tackle the multi modal retrieval problem, performance is currently still far from perfect, and it remains an open question as to how similar the training/test samples should be, since different methods have different tolerances, especially for challenging sketches. We expect to see more research tackling this in the future.

Acknowledgements

We would like to thank Enrique Rosales, Alla Sheffer, and Giorgio Gori for providing the FlowRep example in Figure 5.13.

References

[1] H. F. Alhasson, S. S. Alharbi, and B. Obara. 2D and 3D Vascular Structures Enhancement via Multiscale Fractional Anisotropy Tensor. In *ECCV Workshop: Bioimage Computing*, pages 365–374, 2018.

[2] P. Arbelaez, M. Maire, C. Fowlkes, and J. Malik. Contour detection and hierarchical image segmentation. *IEEE Transactions on Pattern Analysis and Machine Intelligence*, 33(5):898–916, 2010.

[3] P. J. Basser, J. Mattiello, and D. LeBihan. MR Diffusion Tensor Spectroscopy and Imaging. *Biophysical Journal*, 66(1):259–267, 1994.

[4] S. Belongie, J. Malik, and J. Puzicha. Shape Matching And Object Recognition Using Shape Contexts. *IEEE Transactions on Pattern Analysis and Machine Intelligence*, 24(4):509–522, 2002.

[5] P. Bénard and A. Hertzmann. Line Drawings from 3D Models: A Tutorial. *Foundations and Trends in Computer Graphics and Vision*, 11(1-2):1–159, 2019.

[6] J. Canny. A Computational Approach to Edge Detection. *IEEE Transactions on Pattern Analysis and Machine Intelligence*, 8:679–698, 1986.

[7] S. Carsten. An Unbiased Detector of Curvilinear Structures. *IEEE Transactions on Pattern Analysis and Machine Intelligence*, 20(2): 113–125, 1998.

[8] D. Y. Chen, X. P. Tian, Y. T. Shen, and M. Ouhyoung. On Visual Similarity Based 3d Model Retrieval. In *Computer Graphics Forum*, volume 22, pages 223–232. Wiley Online Library, 2003.

[9] F. Cole, A. Golovinskiy, A. Limpaecher, H. S. Barros, A. Finkelstein, T. Funkhouser, and S. Rusinkiewicz. Where Do People Draw Lines? *ACM Transactions on Graphics*, 27(3):1–1, 2008.

[10] W. Dai and S. Liang. Cross-Modal Guidance Network For Sketch-Based 3D Shape Retrieval. In *2020 IEEE International Conference on Multimedia and Expo (ICME)*, pages 1–6. IEEE, 2020.

[11] N. Dalal and B. Triggs. Histograms Of Oriented Gradients For Human Detection. In *2005 IEEE computer society conference on computer vision and pattern recognition (CVPR'05)*, volume 1, pages 886–893. IEEE, 2005.

[12] F. De Goes, S. Goldenstein, M. Desbrun, and L. Velho. Exoskeleton: Curve Network Abstraction for 3D Shapes. *Computers & Graphics*, 35 (1):112–121, 2011.

[13] D. DeCarlo, A. Finkelstein, S. Rusinkiewicz, and A. Santella. Suggestive Contours for Conveying Shape. *ACM Transactions on Computer Graphics*, 22(3):848–855, 2003.

[14] M. Eitz, J. Hays, and M. Alexa. How Do Humans Sketch Objects? *ACM Transactions on graphics (TOG)*, 31(4):1–10, 2012.

[15] M. Eitz, R. Richter, T. Boubekeur, K. Hildebrand, and M. Alexa. Sketch-Based Shape Retrieval. *ACM Transactions on Graphics*, 31(4): 1–10, 2012.

[16] Aly A. Farag and Edward J. Delp. Edge Linking by Sequential Search. *Pattern Recognition*, 28(5):611–633, 1995.

[17] M. Flickner, H. Sawhney, W. Niblack, J. Ashley, Q. Huang, B. Dom, M. Gorkani, J. Hafner, D. Lee, D. Petkovic, D. Steele, and D. Yanker. Query by image and video content: The QBIC system. *Computer*, 28 (9):23–32, 1995.

[18] A. F. Frangi, W. J. Niessen, K. L. Vincken, and M. A. Viergever. Multiscale Vessel Enhancement Filtering. In *International conference on medical image computing and computer-assisted intervention*, pages 130–137, 1998.

[19] S. Frank Y. and C. Shouxian. Adaptive Mathematical Morphology For Edge Linking. *Information Sciences*, 167(1):9–21, 2004.

[20] T. Funkhouser, P. Min, M. Kazhdan, J. Chen, A. Halderman, D. Dobkin, and D. Jacobs. A Search Engine For 3d Models. *ACM Transactions on Graphics (TOG)*, 22(1):83–105, 2003.

[21] T. Furuya and R. Ohbuchi. Dense Sampling And Fast Encoding For 3D Model Retrieval Using Bag-of-Visual Features. In *Proceedings of the ACM international conference on image and video retrieval*, pages 1–8, 2009.

[22] T. Furuya and R. Ohbuchi. Similarity Metric Learning For Sketch-based 3D Object Retrieval. *Multimedia Tools and Applications*, 74(23): 10367–10392, 2015.

[23] C. Gao, M. Tang, X. Liang, Z. Su, and C. Zou. PencilArt: A Chromatic Penciling Style Generation Framework. *Computer Graphics Forum*, 37 (6):395–409, 2018.

[24] L. A. Gatys, A. S. Ecker, and M. Bethge. Image Style Transfer Using Convolutional Neural Networks. In *IEEE Conference on Computer Vision and Pattern Recognition*, pages 2414–2423, 2016.

[25] A. Gehre, I. Lim, and L. Kobbelt. Adapting Feature Curve Networks To A Prescribed Scale. *Computer Graphics Forum*, 35(2):319–330, 2016.

[26] T. Goodwin, I. Vollick, and A. Hertzmann. Isophote Distance: A Shading Approach To Artistic Stroke Thickness. In *Proceedings of the ACM Symposium on Non-photorealistic Animation and Rendering*, pages 53–62, 2007.

[27] G. Gori, A. Sheffer, N. Vining, E. Rosales, N. Carr, and T. Ju. FlowRep: Descriptive Curve Networks for Free-Form Design Shapes. *ACM Transactions on Graphics*, 36(4):1–14, 2017.

[28] St Grabli, E. Turquin, F. Durand, and F. X. Sillion. Programmable Rendering of Line Drawing from 3D Scenes. *ACM Transactions on Graphics*, 29(2), 2010.

[29] Y. Gryaditskaya, M. Sypesteyn, J. W. Hoftijzer, S. Pont, F. Durand, and A. Bousseau. Opensketch: A Richly-annotated Dataset Of Product Design Sketches. *ACM Transactions on Graphics*, 38(6):232, 2019.

[30] P. Isola, J. Y. Zhu, T. Zhou, and A. A. Efros. Image-to-image Translation With Conditional Adversarial Networks. In *IEEE Conference on Computer Vision and Pattern Recognition*, pages 1125–1134, 2017.

[31] T. Judd, F. Durand, and E. H. Adelson. Apparent Ridges For Line Drawing. *ACM Transactions on Computer Graphics*, 26(3):19, 2007.

[32] H. Kang, S. Lee, and C. K. Chui. Coherent Line Drawing. In *Proceedings of the ACM Symposium on Non-photorealistic Animation and Rendering*, pages 43–50, 2007.

[33] M. Kazhdan, T. Funkhouser, and S. Rusinkiewicz. Rotation Invariant Spherical Harmonic Representation Of 3D Shape Descriptors. In *Symposium On Geometry Processing*, volume 6, pages 156–164, 2003.

[34] M. Kolomenkin, I. Shimshoni, and A. Tal. On Edge Detection On Surfaces. In *IEEE Conference on Computer Vision and Pattern Recognition*, pages 2767–2774, 2009.

[35] J. E. Kyprianidis, J. Collomosse, T. Wang, and T. Isenberg. State of the "Art": A Taxonomy of Artistic Stylization Techniques for Images and Video. *IEEE Transactions on Visualization and Computer Graphics*, 19 (5):866–885, 2012.

[36] A. Li, S. Shan, X. Chen, and W. Gao. Maximizing Intra-individual Correlations For Face Recognition Across Pose Differences. In *IEEE Conference on Computer Vision and Pattern Recognition*, pages 605–611, 2009.

[37] B. Li and H. Johan. Sketch-based 3D Model Retrieval by Incorporating 2D-3D Alignment. *Multimedia tools and applications*, 65(3):363–385, 2013.

[38] B. Li, Y. Lu, A. Godil, T. Schreck, B. Bustos, A. Ferreira, T. Furuya, M. J. Fonseca, H. Johan, T. Matsuda, R. Ohbuchi, P. B. Pascoal, and J. M. Saavedra. A Comparison Of Methods For Sketch-based 3D Shape Retrieval. *Computer Vision and Image Understanding*, 119:57–80, 2014.

[39] M. Li, Z. Lin, R. Mech, E. Yumer, and D. Ramanan. Photo-sketching: Inferring Contour Drawings from Images. In *IEEE Winter Conference on Applications of Computer Vision (WACV)*, pages 1403–1412, 2019.

[40] D. G. Lowe. Distinctive Image Features From Scale-invariant Keypoints. *International Journal Of Computer Vision*, 60(2):91–110, 2004.

[41] C. Lu, L. Xu, and J. Jia. Combining Sketch And Tone For Pencil Drawing Production. In *Proc. ACM Symposium on Non-photorealistic Animation and Rendering*, pages 65–73, 2012.

[42] Z. Lu, J. Guo, J. Xiao, Y. Wang, X. Zhang, and D. M. Yan. Extracting Cycle-Aware Feature Curve Networks from 3D Models. *Computer-Aided Design*, 131:102949, 2021.

[43] T. Luft, F. Kobs, W. Zinser, and O. Deussen. Watercolor Illustrations of CAD Data. In *Proceedings of the Fourth Eurographics Conference on Computational Aesthetics in Graphics, Visualization and Imaging*, pages 57–63, 2008.

[44] C. Mingjia, W. Changbo, and L. Ligang. Cross-Domain Retrieving Sketch and Shape Using Cycle CNNs. *Computers & Graphics*, 89:50–58, 2020.

[45] T. H. Nguyen-Phuoc, C. Li, S. Balaban, and Y. Yang. RenderNet: A Deep Convolutional Network for Differentiable Rendering From 3D Shapes. In *Advances in Neural Information Processing Systems*, pages 7891–7901, 2018.

[46] J. P. W. Pluim, J. B. A. Maintz, and M. A. Viergever. Mutual-information-based Registration Of Medical Images: A Survey. *IEEE Transactions on Medical Imaging*, 22(8):986–1004, 2003.

[47] C. Romanengo, S. Biasotti, and B. Falcidieno. Recognising Decorations In Archaeological Finds Through The Analysis Of Characteristic Curves On 3D Models. *Pattern Recognition Letters*, 131:405–412, 2020.

[48] P. L. Rosin and J. Collomosse. *Image and Video-based Artistic Stylisation*. Springer, 2013.

[49] K. Sasaki, S. Iizuka, E. Simo-Serra, and H. Ishikawa. Joint Gap Detection And Inpainting Of Line Drawings. In *Proceedings of the IEEE Conference on Computer Vision and Pattern Recognition (CVPR)*, July 2017.

[50] R. Schmidt, A. Khan, G. Kurtenbach, and K. Singh. On Expert Performance in 3D Curve-Drawing Tasks. In *Proceedings of the EUROGRAPHICS Symposium on Sketch-Based Interfaces and Modeling (SBIM '09)*, pages 133–140, 2009.

[51] T. Shao, W. Xu, K. Yin, J. Wang, K. Zhou, and B. Guo. Discriminative Sketch-based 3D Model Retrieval Via Robust Shape Matching. In *Computer Graphics Forum*, volume 30, pages 2011–2020. Wiley Online Library, 2011.

[52] A. Sharma and D. W. Jacobs. Bypassing Synthesis: PLS For Face Recognition With Pose, Low-resolution And Sketch. In *IEEE Conference on Computer Vision and Pattern Recognition*, pages 593–600, 2011.

[53] P. Shilane, P. Min, M. Kazhdan, and T. Funkhouser. The Princeton Shape Benchmark. In *Proceedings Shape Modeling Applications, 2004.*, pages 167–178. IEEE, 2004.

[54] J. Sivic and A. Zisserman. Video Google: A Text Retrieval Approach To Object Matching In Videos. In *Proceedings Ninth IEEE International Conference on Computer Vision*, page 1470. IEEE, 2003.

[55] J. V. B. Soares, J. J. G. Leandro, R. M. Cesar, H. F. Jelinek, and M. J. Cree. Retinal Vessel Segmentation Using The 2D Gabor Wavelet And Supervised Classification. *IEEE Transactions on Medical Imaging*, 25 (9):1214–1222, 2006.

[56] P. Soille. *Morphological Image Analysis: Principles and Applications.* Springer Science & Business Media, 2013.

[57] H. Su, S. Maji, E. Kalogerakis, and E. Learned-Miller. Multi-view Convolutional Neural Networks For 3D Shape Recognition. In *Proceedings of the IEEE international conference on computer vision*, pages 945–953, 2015.

[58] J. T. Todd. The Visual Perception Of 3D Shape. *Trends In Cognitive Sciences*, 8(3):115–121, 2004.

[59] R. G. von Gioi and G. Randall. Unsupervised Smooth Contour Detection. *Image Processing On Line*, 6:233–267, 2016.

[60] F. Wang, L. Kang, and Y. Li. Sketch-based 3D Shape Retrieval Using Convolutional Neural Networks. In *Proceedings of the IEEE Conference on Computer Vision and Pattern Recognition*, pages 1875–1883, 2015.

[61] X. Wang and X. Tang. Face Photo-sketch Synthesis and Recognition. *IEEE Transactions on Pattern Analysis and Machine Intelligence*, 31 (11):1955–1967, 2008.

[62] T. Wen-Hsiang. Moment-Preserving Thresholding. *Computer Vision, Graphics and Image Processing*, 29:377–393, 1985.

[63] J. Wu, R. R. Martin, P. L. Rosin, X. F. Sun, Y. K. Lai, Y. H. Liu, and C. Wallraven. Use Of Non-Photorealistic Rendering and Photometric Stereo in Making Bas-Reliefs from Photographs. *Graphical Models*, 76 (4):202–213, 2014.

[64] M. Ye, S. Zhou, and H. Fu. DeepShapeSketch: Generating Hand Drawing Sketches from 3D Objects. In *International Joint Conference on Neural Networks*, pages 1–8, 2019.

[65] S. M. Yoon, M. Scherer, T.s Schreck, and A. Kuijper. Sketch-based 3D Model Retrieval Using Diffusion Tensor Fields of Suggestive

Contours. In *Proceedings of the 18th ACM international conference on Multimedia*, pages 193–200, 2010.

[66] D. Zhang and G. Lu. A Comparative Study of Fourier Descriptors for Shape Representation and Retrieval. In *Proceedings of the 5th Asian Conference on Computer Vision*, page 35, 2002.

[67] D. Zhou, O. Bousquet, T. N. Lal, J. Weston, and B. Schölkopf. Learning With Local And Global Consistency. In *Advances In Neural Information Processing Systems*, pages 321–328, 2004.

Part II

3D Sketching

6

Introduction to 3D Sketching

Rahul Arora[1], Mayra Donaji Barrera Machuca[2], Philipp Wacker[3], Daniel Keefe[4], and Johann Habakuk Israel[5]

[1]University of Toronto, Toronto, Canada
[2]Dalhousie University, Canada
[3]RWTH Aachen University, Germany
[4]University of Minnesota, US
[5]Hochschule für Technik und Wirtschaft Berlin,
University of Applied Sciences, Germany

What if designers' marks and movements while sketching could transcend the page and exist in 3D space? This chapter addresses this exciting question through a fresh discussion of the science, techniques, and applications of sketching in 3D space. Although 3D spatial relationships are often depicted in traditional 2D sketches, this chapter focuses specifically on a different form of sketching that may be new to many readers. Here, sketching in 3D space refers to a type of technology-enabled sketching where

1. the physical act of mark-making is accomplished off-the-page in a 3D, body-centric space,
2. a computer-based tracking system records the spatial movement of the drawing implement, and
3. the resulting sketch is often displayed in this same 3D space, for example, via the use of immersive computer displays, as in virtual and augmented realities (VR and AR).

Although such technologies have only recently matured to the point where practical limitations such as costs and maintenance among others are no longer major issues, it is already clear from the early work reviewed here that sketching in 3D space has serious potential to transform product design and other design fields.

To better understand this potential, let us first reflect on the role of some existing 3D design tools, for example, physical prototyping, and compare this to sketching. When we interact with a physical 3D prototype of a product, architectural space, or another design idea, we are able to make body-centric spatial judgments. We can measure lengths from the first person perspective, and understand the form, scale, light, and more. We get to experience the design. Unfortunately, physical prototypes also have some limitations. Some designs are expensive, impossible, or time-consuming to prototype in physical form or at a natural scale, and they are also difficult to edit and annotate.

Traditional 2D sketching, so well documented in previous chapters, provides a complementary tool. Sketching is expressive while also being immediate and easily editable. Sketching enables rapid exploration. We can, for instance, make 20 sketches, throw out 19, and be very happy about how the design process is proceeding. Yet, traditional 2D sketches never quite capture the experience of holding or standing within a physical prototype. We cannot use 2D sketches to make body-centric judgments about the scale and other spatial relationships. Even if a sketch beautifully captures a 3D likeness, it does this only from a single vantage point.

Sketching in 3D space promises to overcome these limitations by combining the best of both worlds. In theory, this leads to a medium that provides designers with both the expressiveness, immediacy, and edit-ability of traditional 2D sketching as well as the body-centered spatial awareness, presence, and multiple perspectives afforded by traditional 3D design tools, such as prototyping.

There are already many sketching in 3D space success stories described in the literature. Artists have transported us to playful virtual worlds [66] and explored new forms of digital 3D sculpture that preserve rather than hide evidence of a real human hand behind the form [24]. Architects have translated their initial 2D design sketches into life-size virtual sketches they can iteratively refine in life-size virtual environments [35]. Scientists have prototyped 3D multivariate data visualizations [53] and even selected bundles of fluid flow in immersive data visualizations by sketching 3D lassos [68]. Engineers have prototyped new medical devices [36].

Today, the underlying technologies that are required to enable applications like these come mostly from the fields of virtual and augmented reality. This chapter introduces the novice reader to the underlying technologies for display and rendering in 3D space in Section 6.1.1, and the spatial tracking and environmental sensing technologies in Section 6.1.2.

In recent years, AR and AR technologies have become more widely available through commercial outlets and consequently, there is wider availability of tools that allow for sketching in 3D space. Suffice to say that if readers have access to a VR or AR headset and its corresponding applications manager, they are likely to find at least one 3D sketching application. These applications provide a great starting point for new users, but designers can benefit from understanding a bit more of the history, science, and use cases for sketching in 3D space.

This chapter, therefore, aims to provide designers with two primary resources. First, in Section 6.1, we trace the origins of 3D sketching, including the underlying technologies that make it possible. Second, in Section 6.2 we present the opportunities and challenges of immersive sketching with a discussion grounded in perceptual and human-computer interaction research. From this discussion, we learn what currently works well, and hence the existing opportunities, as well as what does not work well and hence the remaining challenges in the field. The groundwork laid here is built upon in the subsequent chapters. Chapter 7 documents options and best practices for converting input to geometrical representations, covering the four core topics of tracking, filtering, sampling, and mesh creation. Chapter 8 presents what we believe to be the most complete account of 3D sketching interaction devices and techniques ever assembled, including complete references to research and best practices on sketching sub-tasks, strategies and interactive algorithms for increasing control, and matching interaction techniques to the affordances provided by VR/AR hardware. Finally, Chapter 9 brings all these pieces together to present some compelling applications of 3D sketching, organized according to themes that range from conceptual design and creativity to scientific data visualization.

Since AR and AR hardware has only recently begun to move out of the research lab in the form of commercially successful hardware platforms, most designers today have only been exposed to the concept of 3D sketching through the small set of apps already available in the app stores for the most popular one or two hardware platforms. By covering each of these topics in depth, as informed by the history, current practice, and active research topics within VR/AR, human-computer interaction, visualization, and computer graphics research communities, we hope this chapter and the subsequent chapters on 3D sketching provide designers with an approachable and also uniquely complete account of sketching in 3D space.

6.1 Tracing the Technological Origins of 3D Sketching

In this section, we will dive into the historical development of immersive reality, tracing the roots of the technology that makes immersive 3D sketching possible.

Head-mounted displays were pioneered by Sutherland [70] in 1968. Incidentally, a few years before his work on head-mounted displays (HMDs), Sutherland also invented the first digital sketching system. His Sketchpad [69] system is widely considered to be the precursor of modern sketching tools as well as CAD/CAM systems, and further, was one of the first programs that could be fully controlled by a graphical user interface (GUI). It is interesting to note that immersion and digital sketching share the same roots.

Coming back to immersive realities, Sutherland's pioneering AR system utilized a see-through stereoscopic AR display mounted on a bulky HMD tethered to the ceiling as illustrated in Figure 6.1(a). Objects were rendered on a scanline-based cathode-ray tube (CRT) display which was the leading display technology at the time. Just a glance at contemporary VR devices suggests how far the technology has come. Modern VR headsets such as the Oculus Rift S [54] and Vive Pro [31] weigh less than 500 grams, utilize standardized cables available on any modern PC, and have high-resolution organic light-emitting diode (OLED) displays running at 80–90 frames per second. While Sutherland's device could only display primitive wireframe objects, modern devices can render complex photorealistic scenes with ease. Furthermore, modern VR and AR systems allow six degrees of freedom (DoF) tracking of the HMD as illustrated in Figures 6.1(b) and 6.1(c). That is, both the 3D position and the orientation of the three spatial axes are precisely tracked. Precise and low-latency tracking of the headset is required not just for an enhanced sense of presence in the immersive environment but is essential to prevent disorientation and cybersickness as discussed in Section 6.2.2.2. For interaction, users either utilize similarly tracked controllers, or make use of recently developed algorithms for bare-hand tracking. Section 8.2.2 presents a discussion on the differences between these interaction devices.

Here, we present the technical advancements that have made this transformation possible. The idea is to give a broad overview of the area, as an exhaustive summary of all VR/AR technological advancements is beyond the scope of this book. We categorize these advancements into the following areas.

(a) 1968 HMD   (b) Modern HMD: Oculus Rift S   (c) Modern HMD: HTC Vive Pro

Figure 6.1 Sutherland's pioneering AR HMD [70] was bulky and tethered to the ceiling (a). In contrast, modern HMDs afford freedom of movement and interaction via 6-DoF headset and controller tracking (b, c). Some modern devices such as the Oculus Rift S use inside-out tracking (b), while others like the Vive Pro (c) use outside-in tracking which requires external wall-mounted trackers. © (b) Facebook, and (c) HTC.

6.1.1 Display Technologies and Rendering

Immersive VR environments can be simulated using a stereoscopic head-mounted display, first demonstrated by Sutherland [70], or a multi-projected environment, such as the cave automatic virtual environment (CAVE) [13]. Our focus is on HMDs, the dominant modality for modern VR. In the past few decades, VR HMD designers have gained from the improvement in display technologies, leading from CRTs to flat-panel liquid crystal displays (LCDs) and then on to flat as well as curved displays based on light-emitting diodes (LEDs). Apart from the obvious impact on the quality of virtual environments that can be displayed, these advancements have also reduced the power consumption and weight of HMDs, further contributing to the recent explosion of consumer interest in VR.

Another important factor impacting immersion is the display's field of view (FoV). Human eyes have an approximate FoV of 210° horizontally and 150° vertically [71]. Hardware designers have been chasing an improved immersion via higher FoVs from as early as the 1989 Howlett's Cyberface system [46], however, even modern VR devices typically only achieve 110–130° horizontal FoV [73] falling short of what the real world affords. Nevertheless, technological improvements are constantly being made to further push the FoV boundaries [56].

One final technological advancement that has led to high-resolution displays in modern VR systems is the miniaturization of LCD and LED display machinery. With increasing pixel density, hardware designers are now able to integrate high-resolution screens in compact wearable devices. For example, the Vive Cosmos Elite has a resolution of 1440×1700 pixels per eye [73].

Rendering complex scenes on these high-resolution displays while maintaining acceptable frame rates requires massive amounts of computational power. Immersion also requires realistically simulating lighting and shading based on physical principles which can be daunting even for modern graphics processing units (GPUs). As a result, techniques for focusing computational power for synthesizing the most perceptually-salient regions of the scene is an active area of research. An important technique for VR HMDs is foveated rendering [55], which synthesizes lower details in the periphery compared to the user's point of focus on the *fovea*. While current commercial devices only offer "fixed" foveated rendering, which renders the edges of the display at a lower resolution, mounting eye tracking sensors on HMDs is being actively researched to enable true foveated rendering [81].

Compared to VR, augmented reality displays need to tackle the additional challenge of combining the real world with the virtual. While projection-based AR has been a widely-studied area as well, see for example, [59], we will focus on headset-based and mobile AR. Modern AR devices such as the Magic Leap 1 [51] and Hololens 2 [52] use an optical see-through display. Such display systems include beam-splitters to combine the real-world image with the reflection of an image produced by a stereoscopic display [9]. In contrast, mobile AR is video-based, that is, the technology overlays virtual images over an image of the real world captured through a video camera. While optical see-through AR offers a better sense of immersion, since the user can directly see the physical world, a temporal delay between the real world and virtual objects is always present. This delay occurs because any immersive experience requires sensing and processing the real world and manipulating the virtual world in response. However, the real world can change by the time the virtual objects respond to it, and therefore successfully executing immersive see-through AR requires extremely low-latency sensing hardware and processing algorithms.

6.1.2 Spatial Tracking and Environment Sensing

The second necessary ingredient for realizing an immersive environment is spatial tracking. Six degrees of freedom (DoF), namely three translational DoF for position and three rotational DoF for orientation, are typically tracked by HMDs and handheld controllers. Rolland et al. [64] provide a survey of a variety of magnetic field based sensors that have historically been utilized for 6-DoF tracking for immersive environments. Magnetic tracking is unaffected by occlusion and optical disturbance, but their short range limits its utility and these tracking devices have fallen out of favor in recent years. Another common tracking technique is through inertial tracking, which utilizes inertial measurement unit (IMU) sensors such as accelerometers, gyroscopes, and magnetometers. These IMU sensors are mounted on an HMD or other objects that require tracking, and directly measure linear acceleration and angular orientation. Unfortunately, IMU sensors are susceptible to drift: accumulated error over time [12]. This is especially true for positional tracking, which requires integrating linear acceleration over time to get velocity, which is then integrated to get a change in position. Therefore, VR/AR devices typically use IMU sensors in combination with other sensing techniques [10].

The third tracking method involves computer vision and has been actively researched since the early 1990s [5, 75] and has shown rapid advancement recently [22, 16]. Vision-based tracking can make use of either the infrared (IR) portion of the electromagnetic spectrum, or the visible light portion. IR-based tracking typically requires an external source of IR light, which is then reflected by IR reflectors on the HMD and controllers [75]. This configuration can be reversed as well such that the HMD contains the light source while the external *trackers* are equipped with cameras [61]. Such systems are called inside-looking-out and outside-looking-in, respectively, sometimes shortened to inside-out and outside-in as shown in Figures 6.1(b) and 6.1(c). Inside-out systems capturing visible light can completely do away with external markers by tracking prominent real-world features such as edges, textures, high-level descriptors such as SIFT [49], and features learned via neural-networks [79]. Recently, computer vision algorithms have also been applied for hand-pose tracking and gesture recognition [48], enabling natural bare-handed input in immersive environments.

It should be noted that computer vision-based tracking is typically accurate but suffers from high latency and low update rates. In contrast, inertial and magnetic sensing have low latency and extremely high update rates. As a result, modern devices tend to use a combination of computer vision, inertial, and/or magnetic sensors [25].

In this section, we looked at how display and rendering advancements let us visualize beautiful scenes and models in VR/AR and at the advancements in 3D tracking which allow positioning marks freely in 3D. However, tracking and hardware innovations by themselves do not transform immersive reality into a creative platform. The next section looks at the novel interactions made possible by modern VR/AR and how an artist can harness immersive environments for creative sketching.

6.2 Opportunities and Challenges of Immersive Sketching

The exciting and unprecedented creative potential of immersive realities is driven by developments in two broad domains, namely, technological innovations resulting in the design of high-fidelity hardware, and creative user interface design utilizing the novel interactive capabilities afforded by the said hardware. We looked at the technology driving VR/AR in the previous section; this section discusses the novel interactive affordances enabled by the technology. Our focus here is on broad capabilities; specific research projects and devices are described in more detail in Chapter 8. However, we will not just look into the exciting new creative opportunities of VR/AR, but also delve into the sensorimotor and perceptual challenges that arise during the creative use of these technologies. Following this, we will briefly look at research into the learnability of immersive 3D sketching and modeling, before concluding with a discussion of collaborative 3D creation in immersive realities.

6.2.1 Novel Interaction Capabilities and Creative Avenues

As noted earlier, artists and designers have traditionally sketched on a 2D surface using either a physical sheet of paper or a drawing tablet. The resulting stroke-marks are then displayed on a 2D surface either on the sheet of paper itself or on a digital screen rendering the stroke, respectively. In other words, in traditional sketching systems, both the input *creation*) and the output *visualization* is in the two-dimensional domain. Immersive environments fundamentally transform the creation as well as the visualization process, lifting both to the third dimension. This is especially relevant for designing 3D objects meant to be physically fabricated and used in the real world.

   Three-dimensional input functionality is enabled by tracked controllers, allowing designers to forgo the mental projection from 3D to 2D and directly execute 3D strokes mid-air. As a result, designers do not need to worry about sketching aids such as perspective grids and scaffolding [19, Ch. 2] as illustrated in Figures 6.2(a) and 6.2(b). Moreover, a single 3D sketch can convey the full geometric details of the designed object. This is unlike 2D sketching, where a single sketch only depicts the shape as seen from a

(a) Need for scaffolds   (b) Challenging geometry   (c) Lack of scale

Figure 6.2 Working in 2D often requires sketching aids such as scaffolds and perspective grids (a, b), complex objects are difficult to describe in a single 2D sketch (b), and the medium lacks an immersive sense of scale which can benefit the design and illustration of large architectural structures (c). © Rahul Arora (CC BY 4.0), NASA/JPL-Caltech (free to use), and ArtTower (free to use).

particular viewpoint as illustrated in Figure 6.2(b), and depicting complex shapes often requires multiple sketches from different viewpoints [63, Ch. 6]. Another advantage of 3D sketches is their inherent spatiality. For instance, 3D sketches provide a designer with an opportunity to use their own body to assess the scale of the sketch, thereby allowing them to immediately perceive the spatial impact of their designs [33].

While heightened spatial awareness of the designed object for professional designers is important, an even greater potential for impact is communication with relatively untrained stakeholders with lower levels of spatial ability. That is, designers often need to communicate their ideas with non designer peers and end clients, who may not have had the same training. Immersive displays also improve this communication process, since users without a design background no longer have to take the mental leap of interpreting a 3D shape from purely 2D information [39]. Thus, 3D sketches lower the barrier to entry for understanding early stage designs and those untrained in spatial thinking can also gain a faithful understanding of the visualized concept [57]. At the same time, immersive sketching can reduce the designer's effort required to communicate concepts to clients since even loosely drawn ideation sketches can potentially be communicated to the end client.

The third benefit of immersive sketching that we will look into is *scale*. Immersive environments provide the designer with a potentially-infinite 3D canvas to draw in, enabling the drawing of large objects in real-world scale as illustrated in Figure 6.2(c). For example, a furniture designer can draw a table in a 1:1 scale, instead of drawing a vastly scaled-down version on a sheet of paper. Designers of virtual worlds for games and movies can conceptualize the whole environment in scale, judging proportions with respect to their own bodies, or they can immerse themselves in a virtual space such as a car or a specific room. Design-focused commercial tools such as Gravity Sketch [23] and Shapes XR [72] equip users with readily-accessible measurement tools to aid their sense of real-world scale. Immersion can thus help not just in assessing the visual and aesthetic aspects of a design, but its functional aspects as well [2]. For instance, questions about a new car seat design fitting tall drivers or the reachability of a new steering wheel design for short drivers can be answered with higher confidence when designing in context using an immersive environment.

See-through augmented reality can take designing in context a step further by allowing designers to draw *in situ*, placing the drawing in the context of the physical world. For example, interior designers can decorate a real-world room, tools can be sketched over the user's hands, and virtual buildings can be positioned around existing buildings in a city [40, 4]. To some extent, this novel creative capability is also supported by video-based AR on mobile devices and "VR" devices which allow a pass-through mode via an HMD-mounted camera. Access to the real world is useful not just for drawing *around* physical objects, but also for drawing *on* and *with* them. That is, physical objects can also be employed as constraints for anchoring strokes [4] or as props [34], respectively. Edges, contours, and textures in the real world can also act as visual guidance for sketches [74].

Despite numerous advantages and novel creative avenues opened up by sketching in 3D, it also comes with its own set of issues. We look at the most important challenges in the next section.

6.2.2 Challenges in Control and Perception

Creating directly in 3D presents a host of novel challenges, which we divide broadly into three categories, namely issues related to control and precision of 3D strokes, challenges in perceiving objects in stereoscopic 3D, and ergonomic problems encountered by users of 3D sketching systems.

6.2.2.1 Control and Precision Issues

In traditional 2D sketching, a drawing surface provides a physical constraint which helps artists anchor their strokes as shown in Figure 6.3(a). Lacking such a physical constraint, mid-air 3D drawing can be difficult to control and prone to inaccuracy [3, 37]. Keefe et al. [37, 38] indicate that haptic feedback can be useful for reducing control errors, but the range and form-factor of current haptic devices can be limiting. Arora et al. [3] further suggest that 3D drawing imprecision is not limited to out of plane meandering, that is, attempts to draw specific planar stroke mid-air were observed to be less accurate than one drawn on a physical surface even when the mid-air stroke was projected onto the intended drawing plane. The 3D drawing inaccuracies are also affected by the orientation of drawn strokes with respect to the user, with strokes in the fronto-parallel plane exhibiting the least inaccuracies, while the depth axis is the hardest to sketch precisely [37, 7]. Furthermore, while 3D sketching allows the creation of non planar or *space* curves in a single step, as shown in Figure 6.3(b), such curves tend to exhibit even higher levels of inaccuracy [3]. One may argue that this deficiency is not important since CAD is dominated by planar curves and non planar curves are rarely utilized [67, 80]. However, we hypothesize that the proliferation of planar curves in CAD is partially due to limitations imposed by the traditional 2D devices. Therefore, it is important to build novel tools to improve the non planar curve creation workflow in immersive systems.

Follow-up studies [4, 74] indicate that even sketching directly over physical objects in AR can be prone to precision issues. Drawing over real-world objects is useful for conceptualizing decorations and augmentations for those objects. In these cases, sketching precisely is difficult when the object's surface has high curvature regions as illustrated in Figure 6.3(c), unwieldy surface texture, or when it is a fixed or difficult-to-manipulate object, thus forcing the designer to draw in uncomfortable orientations [3, 4, 74]. Lastly, sketching in the context of the real-world can be challenging if the strokes are either too large or too small and there is a sweet spot for the size of 3D strokes to ensure accuracy. Specifically, consider an example scenario of drawing strokes larger than a typical human's arm span. This requires the user to move their whole body while drawing, making it extremely difficult to control the stroke. Even drawing long straight lines becomes difficult as the stroke follows the natural arc of the human arm. On the other hand, small 3D strokes, of the order of a centimeter, are visibly impacted by the jitter caused by the lack of a physical sketching surface.

Section 8.1.1 discusses recent research that attempts to alleviate these challenges by taking cues from the 2D sketching domain, while filtering mechanisms for utilizing imprecise 3D inputs are described in Section 7.1.

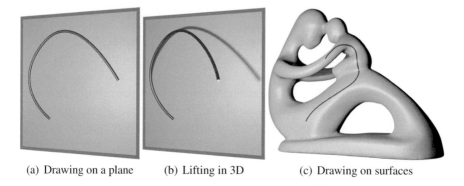

(a) Drawing on a plane   (b) Lifting in 3D   (c) Drawing on surfaces

Figure 6.3 Traditional 2D sketching is limited to a drawing plane (a), while immersive 3D sketching allows lifting marks off the plane to directly create non-planar curves (b), including drawing directly over highly curved physical or virtual objects (c). However, navigating high curvature regions and unusual drawing orientations imposed by such curves can make the execution ergonomically challenging. Fertility model (c) courtesy Aim@Shape repository.

6.2.2.2 Visual and Perceptual Factors

In the real world, humans perceive depth using a variety of perceptual cues, including monocular cues such as lighting, shading, occlusion, and defocus blur; binocular cues such as stereopsis, proprioceptive cues of accommodation and convergence; and dynamic cues such as motion parallax [30]. Virtual objects rendered by HMDs only offer a subset of these cues and the depth perception is, therefore, diminished as compared to the real world. For example, stereopsis and motion parallax is achieved near-trivially by modern hardware due to the presence of stereoscopic displays and stereo rendering. Occlusion can also be easily achieved in VR via depth-ordering techniques, at least for opaque objects. However, the quality of other depth percepts such as lighting and shading may be contingent on the availability of computational resources for rendering the virtual environment in real-time.

In AR environments, occlusion can be also be challenging since correctly occluding and disoccluding the real world requires precise environment sensing. Unfortunately, currently available AR devices suffer from limitations such as limited sensing precision, large initialization lead times, and wildly inaccurate shape estimation when scanning specular (shiny) surfaces.

In some instances, both AR and VR immersive environments can even offer conflicting depth percept cues, causing user discomfort [21]. One important and well-studied problem is the *vergence-accommodation conflict*, caused by the mismatch between the depth indicated by the rendered objects, and the actual depth of the displays [28]. While the eyes accommodate to the screen located at a distance of a centimeter or from the user, they converge onto the distance indicated by the rendered objects, which is often much larger.

An important characteristic of depth cues is their distance dependence [14]. Notably, occlusion is an extremely relative depth discrimination cue for nearby objects (0–2 m away). Since typical 3D sketching modalities involve sketching in the user's vicinity, this suggests that correctly inferring occlusions and disocclusions is an important puzzle to solve for a truly immersive AR sketching experience.

An additional challenge is the dominance of curves as the preferred geometric modeling primitives for creative sketching which we discuss in Chapter 7. Perceiving depth in the real-world typically involves surfaces, which convey additional depth cues via lighting, shading, and textures. Unlike 3D surfaces, thin curves cannot effectively communicate depth to a viewer via these cues. Furthermore, our focus is on an artist's workflow: an artist must be able to precisely position their head and hands in relation to existing strokes to continue drawing accurately. Unfortunately, as demonstrated by Lubos et al. [50], even reaching out to precise positions in three-dimensions can be challenging. Their study involved participants trying to reach out and select flat-shaded disks in 3D, where, flat shading implies that the discs could not convey texture and shading details. The results of their experiment show that errors due to sensorimotor inaccuracies in 3D hand positioning are a minor factor in comparison to errors due to depth misperception [50]. Recent works [6, 8, 3, 37] further characterize 3D selection inaccuracy, demonstrating that the dominant inaccuracy is along the depth axis. For further details on perceptual issues in immersive systems, especially in AR, we refer the reader to the excellent ontology by Drascic and Milgram [18].

Using novel algorithms, interfaces, and hardware design, various remedies for these perceptual issues have been suggested in the literature. We will look into these solutions in Section 8.2.2 and Section 8.2.2.2.

6.2.2.3 Ergonomics and User Comfort

In Section 6.2.2.1, we noted that the physical drawing surface in traditional 2D sketching provides a physical constraint that mid-air sketching lacks. This lack of constraint does not just impact stroke precision, but the removal of a physical surface to support the user's hand and arm also increases the strain on the user's muscles [42, 82]. Furthermore, techniques have been developed to minimize strain when drawing in 2D. For example, artists are taught to draw using larger muscles controlling the shoulder and elbow joints rather than smaller muscles controlling the wrist [63, Ch. 1]. Formal sketching guidelines are yet to be developed for mid-air sketching, and it is unclear if it is even possible to sketch in three-dimensions with the same level of efficiency and comfort that artists currently enjoy in 2D. Still, some generic guidelines for VR design such as prototyping interfaces in VR, embracing the 3D interaction space while still maintaining familiar user interface metaphors, and designing with ergonomics in mind, have started to appear [32].

Numerous lab studies and artist interviews have indicated that fatigue remains an issue in mid-air sketching and modeling [3, 37, 17, 43]. Professional artists state that spending long hours sketching in VR can induce neck and shoulder pain [3]. Experiments also suggest that traditional drawing plane orientations utilized by draftsmen and digital artists are close to optimal, and that mid-air drawing fatigue can be partially mitigated by sketching tools that allow users to draw in these desired orientations by, for example, providing methods for efficiently repositioning the scene [3].

A number of solutions have been suggested to mitigate this added fatigue. A haptic rendering device has been shown to deduce mid-air drawing fatigue [37]. Unfortunately, the range of current commercially-available haptic rendering devices can be extremely limiting, supporting only a distance of around a fifth of a meter or less [1], thereby destroying the large-scale drawing affordance of mid-air sketching. It must be noted, however, that wireless devices do exist as research prototypes as in Section 8.2.2.2. Utilizing a graphic tablet for sketching while mapping the strokes to an immersive visualization space is another solution [4, 17]. However, the need for manipulating a tablet in 3D space, or holding it still in space can also be a demanding task. Even holding a mobile device steadily mid-air can induce fatigue [43]. Finally, Arora et al. [3] suggest that artists should reposition the scene in order to draw in certain desirable orientations, called *sketchable* orientations. Interestingly, the suggested optimal orientations are similar to those utilized by traditional draftsmen. We will return to this discussion in Section 8.2.2.2 and discuss some ideas for improving the ergonomics of immersive sketching in detail.

Another important consideration is the strain induced by the HMD itself. In an immersive environment, *the sensory conflict* caused by the mismatch between the simulated reality and our expectation of reality can induce simulator sickness [60]. In VR, this can be caused, for example, by the disruption of well-trained interaction between senses due to the low refresh rates of the immersive display, as compared to the expectation of the human brain [45]. In AR, *swimming* artifacts caused by poor environment sensing can induce sickness [29]. Fortunately, both rendering and sensing have been rapidly improving, raising the hope that VR/AR-sickness can be largely avoided. For a more thorough discussion on VR-induced cybersickness, see Davis et al. [15]. It must be noted that while the *sensory conflict* theory is the most widely accepted explanation for VR-induced simulator sickness, competing theories suggest the lack of a *rest frame* aligned with the user's inertial frame of reference [58, 76] and *postural instability* caused by the body's attempts to learn to stabilize in a novel virtual environment [62] are also being actively researched.

Lastly, while lighter than their historical counterparts, the weight of immersive reality HMDs still causes a significant strain on the user's head and neck. Innovations in headset design and computational miniaturization can help combat this problem.

6.2.3 Learnability Considerations

In Section 6.2.2.1, we talked about studies that show that humans do not enjoy the same degree of control and precision at 3D sketching as they do at traditional 2D sketching. Even experienced artists encounter a host of difficulties when creating in immersive 3D. However, an important point to note is that these *experienced* artists are experienced in traditional 2D sketching, or other forms of visual creation which have existed for a long time. In contrast, mid-air 3D sketching is a novel medium and most artists have not been exposed to it before. Therefore, the question of the *learnability* of 3D sketching is of paramount importance.

As a species, humans have been sketching in 2D for millennia [47]. Every professional designer has had years of formal training and experience sketching in two-dimensions. Is it fair then, to compare human sketching abilities in the novel mid-air 3D domain to the long-learned 2D sketching? Of more practical importance is the question of training artists to get better at 3D sketching and understanding how quickly this skill can be acquired.

The learnability of immersive sketching was first studied by Wiese et al. [77]. In a study with 25 design students as participants, they observed that even short training sessions of 10–30 minutes produced an observable improvement in 3D sketching quality. Barrera Machuca et al. [7] noted that higher spatial ability, as measured via standard tests [41, 20], was correlated with better 3D shape depiction when using immersive sketching. This suggests that the spatial reasoning and imagination skills acquired by artists and designers during formal training and practice can transfer over to 3D sketching as well.

Much work is still needed to gain a deeper understanding of 3D sketching learnability. Not many learnability studies have been performed so far, and even the ones that have been performed only study participants over short periods of time. Only a long term study that looks at participants' progress over weeks or months can reveal the true potential of 3D sketching. With the learnability considerations in mind, we now move on to a discussion of collaborative creation in immersive realities.

6.2.4 Considerations for Collaborative Creation

Design rarely happens in isolation the synthesis of a useful real-world object requires designers to collaborate with peers, engineers, managers, and clients to brainstorm ideas, offer and receive feedback, and ideate iteratively. It is, therefore, crucial to examine the collaborative aspects of immersive sketching. How can designers communicate sketches created in VR/AR? How does immersion aid this communication? What are the hindrances to a successful collaboration in immersive sketching? In this section, we will look at a few interactive tools designed to aid collaboration in immersive design. We shall also identify the challenges these tools aim to solve, and how the target applications dictate which collaboration tools users desire.

In the study by Herman and Hutka [27], expert 2D artists were introduced to 3D creation in VR to observe how they adapt to the novel medium. They noted artists' expectations for the interface, functionality, and applications, as well as their mental models for creation in 2D and VR. While artists also imagined using VR for creating initial mock-ups and for collaborating with 3D artists who build 3D models based on their 2D concept art, communicating the design to end-clients was considered the most important use case for VR. As we have noted earlier, immersion improves spatial awareness for designers and non-designers alike but has a larger impact on the spatial awareness of non designers. Therefore, artists in Herman and Hutka's study found VR extremely useful for communicating with clients, who do not have formal training in design. Experimental research has confirmed that immersive environments can improve spatial awareness [44] and aid non

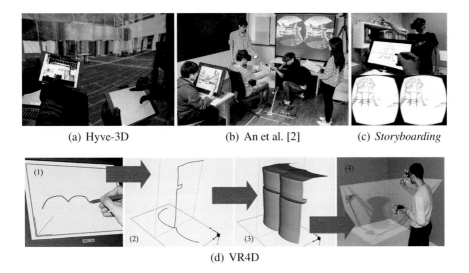

(a) Hyve-3D   (b) An et al. [2]   (c) *Storyboarding*

(d) VR4D

Figure 6.4   Collaborative design using immersive systems. In Hyve-3D [17], individual tablets act as 3D cursors for interacting with shared immersive design space. An et al. [2] target the prototyping of functional user experiences (b). Henrikson et al. [26] support asymmetric collaboration for storyboarding 360° movies (c). VR4D [11] connects a VR user to a tablet user for collaborative sketch-based CAD (d).

professional users in comprehending 3D content [65], thus aiding designer to client communication.

Other studies have looked more closely at designers actively collaborating in immersive settings for applications ranging from architectural and interior design [17, 11] to experience prototyping for the design of automotive interiors [2] to storyboarding for VR movies [26].

To aid designer collaboration for architectural design, Dorta et al. [17] created the Hyve-3D system, which equips designers with individual tablets for interacting with a shared anamorphic projection-based immersive 3D space for architectural design. As shown in Figure 6.4(a), the tablet acts as an individual *3D cursor*, allowing users a personal window for visualization as well as creation. The individual tablets allow users to co-design while optionally looking at the shared immersive visualization for communication. Unfortunately, Hyve-3D does not allow mid-air sketching or any other technique for creating non-planar geometry, as all strokes are executed on the tablet and are constrained to be planar.

VR4D [11] explores the collaboration between users who do not share an immersive space. In their system, one user dons a VR HMD, while the other holds a drawing tablet. As shown in Figure 6.4(d), collaboration is enabled by dividing the design duties between the users. While the tablet user creates geometry by **sketching** 2D curves (1) and building surfaces of extrusion and revolution (3), the immersed user **reviews** the created geometry (2), provides verbal **feedback**, and handles scene **layout** (4). In addition to the shared workspace, VR4D aids collaboration by rendering a rectangle representing the tablet user's drawing plane in VR, and by relaying the sketched strokes to the VR user in real-time. Henrikson et al. [26] employ similar metaphors for 360° storyboard design. In their system, the artist holds the tablet, while the director wears the HMD as shown in Figure 6.4(c). Collaboration aids include real-time visualization of the artist's strokes for the director, maintaining a consistent FoV across both modalities, a shared ground plane with a radial grid to aid verbal communication, and overlays indicating the current view for both users. Furthermore, when receiving feedback from the director, the artist can choose to couple the two views, allowing the director's HMD to dictate the tablet's viewport as illustrated in Figure 6.4(c)(bottom).

An et al. [2] take this metaphor a step forward by imagining multiple members of a design team in a collaborative design setting. In their study, designers are provided with a barrage of VR and AR headsets, drawing tablets, and physical props as shown in Figure 6.4(b). Team-members play different roles and utilize the device or prop most suitable for their assigned role. The studied design task is to experience prototyping of automotive interior interfaces. Using a combination of shared and isolated experiences, the designers collaborate to design the shape of the car interiors and the functionality of the interface.

The lesson to takeaway from these case studies is that while VR provides novel opportunities for collaboration, specific design domains need careful treatment for their particular demands. However, some general solutions can potentially be applied across domains. For example, Xia et al. [78] recently developed a set of techniques for helping resolve spatiotemporal conflicts in virtual scene design in a collaborative immersive setting. This system helps reduce conflicts by letting users work on parallel copies of scene objects, and aiding collaborative interactions among variably-sized *avatars* corresponding to individual users. It will be interesting to adapt such techniques for 3D design sketching.

6.3 Summary

In this chapter, we delved into the historical background of 3D sketching, looking into the technology underpinning contemporary 3D sketching hardware. We further discussed the novel creative opportunities enabled by this exciting medium and how researchers are designing novel tools to exploit these opportunities. But we also discussed the flip side of the coin—the challenges in control and precision, ergonomics, and perception which can hinder creativity in immersive environments. Fortunately, significant research has already gone into novel interaction techniques and input devices that help mitigate these problems, and are comprehensively described in Chapter 8. But before describing these solutions, the next chapter further describes mechanisms for processing 3D sketch inputs, further laying the groundwork that will allow the reader to understand existing 3D sketching systems, modify and augment them, and even build their own 3D sketching tools.

References

[1] 3D Systems. Touch haptic device technical specifications. https://www.3dsystems.com/haptics-devices/touch/specifications, 2020.

[2] S. G. An, Y. Kim, J. H. Lee, and S. H. Bae. Collaborative Experience Prototyping of Automotive Interior in VR with 3D Sketching and Haptic Helpers. In *Proceedings of the 9th International Conference on Automotive User Interfaces and Interactive Vehicular Applications*, AutomotiveUI '17, page 183–192, 2017.

[3] R. Arora, R. H. Kazi, F. Anderson, T. Grossman, K. Singh, and G. Fitzmaurice. Experimental Evaluation of Sketching on Surfaces in VR. In *Proceedings of the 2017 CHI Conference on Human Factors in Computing Systems*, CHI '17, page 5643–5654, 2017.

[4] R. Arora, R. Habib K., T. Grossman, G. Fitzmaurice, and K. Singh. Symbiosissketch: Combining 2D & 3D Sketching for Designing Detailed 3D Objects in Situ. In *Proceedings of the 2018 CHI Conference on Human Factors in Computing Systems*, pages 1–15, 2018.

[5] R. Azuma and G. Bishop. Improving Static and Dynamic Registration in an Optical See-through HMD. In *Proceedings of the 21st Annual Conference on Computer Graphics and Interactive Techniques*, SIGGRAPH '94, page 197–204, 1994.

[6] M. D. Barrera Machuca and W. Stuerzlinger. The Effect of Stereo Display Deficiencies on Virtual Hand Pointing. In *Proceedings of the 2019 CHI Conference on Human Factors in Computing Systems*, CHI '19, 2019.

[7] M. D. Barrera Machuca, W. Stuerzlinger, and Paul Asente. The Effect of Spatial Ability on Immersive 3D Drawing. In *Proceedings of the 2019 on Creativity and Cognition*, C&C '19, page 173–186, 2019.

[8] A. U. Batmaz, M. D. Barrera Machuca, D. M. Pham, and W. Stuerzlinger. Do Head-Mounted Display Stereo Deficiencies Affect 3D Pointing Tasks in AR and VR? In *2019 IEEE Conference on Virtual Reality and 3D User Interfaces (VR)*, pages 585–592. IEEE, 2019.

[9] M. Billinghurst, A. Clark, and G. Lee. A Survey Of Augmented Reality. *Foundations and Trends® in Human–Computer Interaction*, 8(2-3): 73–272, 2015.

[10] G. Bleser and D. Stricker. Advanced Tracking Through Efficient Image Processing And Visual–inertial Sensor Fusion. *Computers & Graphics*, 33(1):59–72, 2009.

[11] A. Chellali, F. Jourdan, and C. Dumas. VR4D: An Immersive and Collaborative Experience to Improve the Interior Design Process. In *5th Joint Virtual Reality Conference of EGVE and EuroVR, JVRC 2013*, pages 61–65, 2013.

[12] C. Chen, X. Lu, A. Markham, and N. Trigoni. Ionet: Learning To Cure The Curse Of Drift In Inertial Odometry. In *Thirty-Second AAAI Conference on Artificial Intelligence*, 2018.

[13] C. Cruz-Neira, D. J. Sandin, T. A. DeFanti, R. V. Kenyon, and J. C. Hart. The CAVE: Audio Visual Experience Automatic Virtual Environment. *Communications of the ACM*, 35(6):64–72, 1992.

[14] J. E. Cutting and P. M. Vishton. Perceiving Layout and Knowing Distances: The Integration, Relative Potency, and Contextual use of Different Information About Depth. In *Perception of space and motion*, pages 69–117. Elsevier, 1995.

[15] S. Davis, K. Nesbitt, and E. Nalivaiko. A Systematic Review Of Cybersickness. In *Proceedings of the 2014 Conference on Interactive Entertainment*, pages 1–9, 2014.

[16] D. DeTone, T. Malisiewicz, and A. Rabinovich. SuperPoint: Self-Supervised Interest Point Detection and Description. In *The IEEE Conference on Computer Vision and Pattern Recognition (CVPR) Workshops*, June 2018.

[17] T. Dorta, G. Kinayoglu, and M. Hoffmann. Hyve-3D and the 3D Cursor: Architectural co-design with freedom in Virtual Reality. *International Journal of Architectural Computing*, 14(2):87–102, 2016.

[18] D. Drascic and P. Milgram. Perceptual Issues In Augmented Reality. In *SPIE Vol. 2653: Stereoscopic Displays and Virtual Reality Systems III*, volume 2653, pages 123–134, San Jose, 1996.

[19] K. Eissen and R. Steur. *Sketching: Drawing Techniques for Product Designers*. Laurence King Publishing, 2007.

[20] R. B. Ekstrom, D. Dermen, and H. H. Harman. *Manual for Kit of Factor-Referenced Cognitive Tests*, volume 102. Educational testing service Princeton, NJ, 1976.

[21] F. El Jamiy and R. Marsh. Survey On Depth Perception In Head Mounted Displays: Distance Estimation In Virtual Reality, Augmented Reality, And Mixed Reality. *IET Image Processing*, 13(5):707–712, 2019.

[22] M. Garon and J. F. Lalonde. Deep 6-DOF Tracking. *IEEE transactions on visualization and computer graphics*, 23(11):2410–2418, 2017.

[23] Gravity Sketch. Gravity Sketch. https://www.gravitysketch.com/, 2020.

[24] J. Grey. Human-Computer Interaction In Life Drawing, A Fine Artist's Perspective. In *Proceedings of the Sixth International Conference on Information Visualisation*, pages 761–770, London, UK, 2002. IEEE Comput. Soc.

[25] C. He, P. Kazanzides, H. T. Sen, S. Kim, and Y. Liu. An Inertial And Optical Sensor Fusion Approach For Six Degree-of-freedom Pose Estimation. *Sensors*, 15(7):16448–16465, 2015.

[26] R. Henrikson, B. Araujo, F. Chevalier, K. Singh, and R. Balakrishnan. Multi-Device Storyboards for Cinematic Narratives in VR. In *Proceedings of the 29th Annual Symposium on User Interface Software and Technology - UIST '16*, pages 787–796, Tokyo, Japan, 2016.

[27] L. M. Herman and S. Hutka. Virtual Artistry: Virtual Reality Translations of Two-Dimensional Creativity. In *Proceedings of the 2019 on Creativity and Cognition*, C&C '19, page 612–618, 2019.

[28] D. M. Hoffman, A. R. Girshick, K. Akeley, and M. S. Banks. Vergence–accommodation Conflicts Hinder Visual Performance And Cause Visual Fatigue. *Journal of Vision*, 8(3):33–33, 03 2008.

[29] R. L. Holloway. *Registration Errors in Augmented Reality Systems*. PhD thesis, University of North Carolina at Chapel Hill, USA, 1996.

[30] I. P. Howard and B. J. Rogers. *Seeing In Depth, Vol. 2: Depth Perception.* University of Toronto Press, 2002.

[31] HTC. Vive PRO | The Professional Grade VR Headset. https://www.vi ve.com/eu/product/vive-pro/, 2018.

[32] Blake Hudelson. Designing for VR: A beginners guide, March 2017. URL https://blog.marvelapp.com/designing-vr-beginners-guide/. [Online; accessed 2020-07-10].

[33] J. H. Israel, E. Wiese, M. Mateescu, C. Zöllner, and R. Stark. Investigating Three-dimensional Sketching For Early Conceptual Design—results From Expert Discussions And User Studies. *Computers & Graphics*, 33(4):462 – 473, 2009.

[34] B. Jackson and D. F. Keefe. Sketching Over Props: Understanding and Interpreting 3D Sketch Input Relative to Rapid Prototype Props. In *IUI 2011 Sketch Recognition Workshop*, 2011.

[35] B. Jackson and D. F. Keefe. Lift-Off: Using Reference Imagery and Freehand Sketching to Create 3D Models in VR. *IEEE Transactions on Visualization and Computer Graphics*, 22(4):1442–1451, 2016.

[36] Seth Johnson, Bret Jackson, Bethany Tourek, Marcos Molina, Arthur G. Erdman, and Daniel F. Keefe. Immersive analytics for medicine: Hybrid 2d/3d sketch-based interfaces for annotating medical data and designing medical devices. In *Proceedings of the 2016 ACM Companion on Interactive Surfaces and Spaces*, ISS '16 Companion, page 107–113, 2016.

[37] D. Keefe, R. Zeleznik, and D. Laidlaw. Drawing on Air: Input Techniques for Controlled 3D line Illustration. *IEEE Transactions on Visualization and Computer Graphics*, 13(5):1067–1081, 2007.

[38] D. F. Keefe, D. Acevedo, J. Miles, F. Drury, S. M. Swartz, and D. H. Laidlaw. Scientific Sketching for Collaborative VR Visualization Design. *IEEE Transactions on Visualization and Computer Graphics*, 14(4):835–847, July 2008.

[39] D. Keeley. The Use Of Virtual Reality Sketching In The Conceptual Stages Of Product Design. Master's thesis, Bournemouth University, 2018.

[40] Y. Kim and S. H. Bae. SketchingWithHands: 3D Sketching Handheld Products with First-Person Hand Posture. In *Proceedings of the 29th Annual Symposium on User Interface Software and Technology*, pages 797–808, 2016.

[41] M. Kozhevnikov and M. Hegarty. A Dissociation Between Object Manipulation Spatial Ability And Spatial Orientation Ability. *Memory & Cognition*, 29(5):745–756, 2001.

[42] P. G. Kry, A. Pihuit, A. Bernhardt, and M. P. Cani. Handnavigator: Hands-on Interaction For Desktop Virtual Reality. In *Proceedings of the 2008 ACM symposium on Virtual reality software and technology*, pages 53–60, 2008.

[43] K. C. Kwan and H. Fu. Mobi3dsketch: 3D Sketching in Mobile AR. In *Proceedings of the 2019 CHI Conference on Human Factors in Computing Systems*, pages 1–11, 2019.

[44] J. J. LaViola, E. Kruijff, R. P. McMahan, D. Bowman, and I. P. Poupyrev. *3D User Interfaces: Theory and Practice*. Addison-Wesley, Boston, 2 edition, 4 2017.

[45] J. J. LaViola Jr. A Discussion Of Cybersickness In Virtual Environments. *ACM SIGCHI Bulletin*, 32(1):47–56, 2000.

[46] LeepVR. The original cyberface and the LEEPvideo system 1. http://www.leepvr.com/cyberface1.php, 1989.

[47] A. Leroi-Gourhan and E. Anati. *The Dawn of European Art: An Introduction to Palaeolithic Cave Painting*. Cambridge University Press Cambridge, 1982.

[48] H. Liang, J. Yuan, D. Thalmann, and N. M. Thalmann. AR in Hand: Egocentric Palm Pose Tracking and Gesture Recognition for

Augmented Reality Applications. In *Proceedings of the 23rd ACM International Conference on Multimedia*, MM '15, page 743–744, 2015.

[49] D. G. Lowe. Distinctive Image Features From Scale-invariant Keypoints. *International Journal Of Computer Vision*, 60(2):91–110, 2004.

[50] P. Lubos, G. Bruder, and F. Steinicke. Analysis Of Direct Selection In Head-mounted Display Environments. In *2014 IEEE Symposium on 3D User Interfaces (3DUI)*, pages 11–18, 2014.

[51] Magic Leap. Magic Leap 1. https://www.magicleap.com/en-us/magic-leap-1/, 2019.

[52] Microsoft. HoloLens 2 | Mixed Reality Technology for Business. https://www.microsoft.com/en-us/hololens/, 2019.

[53] J. Novotny, J. Tveite, M. L. Turner, S. Gatesy, F. Drury, P. Falkingham, and D. H. Laidlaw. Developing Virtual Reality Visualizations for Unsteady Flow Analysis of Dinosaur Track Formation using Scientific Sketching. *IEEE Transactions on Visualization and Computer Graphics*, 25(5):2145–2154, May 2019.

[54] Oculus by Facebook. Oculus Rift S: VR Headset for VR Ready PCs. https://www.oculus.com/rift-s/, 2019.

[55] A. Patney, M. Salvi, J. Kim, A. Kaplanyan, C. Wyman, N. Benty, D. Luebke, and A. Lefohn. Towards Foveated Rendering for Gaze-Tracked Virtual Reality. *ACM Transactions on Graphics*, 35(6), November 2016.

[56] Pimax. Vision 8K PLUS. https://www.pimax.com/products/vision-8k-plus-withoutmas, 2020.

[57] M. Pittalis and C. Christou. Types Of Reasoning In 3D Geometry Thinking And Their Relation With Spatial Ability. *Educational Studies in Mathematics*, 75(2):191–212, 2010.

[58] J. D. Prothero, M. H. Draper, T. A. Furness 3rd, D. E. Parker, and M. J. Wells. The Use Of An Independent Visual Background To Reduce Simulator Side-effects. *Aviation, space, and environmental medicine*, 70 (3 Pt 1):277–283, 1999.

[59] R. Raskar, G. Welch, M. Cutts, A. Lake, L. Stesin, and H. Fuchs. The Office of the Future: A Unified Approach to Image-Based Modeling and Spatially Immersive Displays. In *Proceedings of the 25th*

Annual Conference on Computer Graphics and Interactive Techniques, SIGGRAPH '98, page 179–188, 1998.

[60] J. T. Reason and J. J. Brand. *Motion Sickness*. Academic Press, 1975.

[61] M. Ribo, A. Pinz, and A. L. Fuhrmann. A New Optical Tracking System For Virtual And Augmented Reality Applications. In *IMTC 2001. Proceedings of the 18th IEEE Instrumentation and Measurement Technology Conference. Rediscovering Measurement in the Age of Informatics*, volume 3, pages 1932–1936. IEEE, 2001.

[62] G. E. Riccio and T. A. Stoffregen. An Ecological Theory of Motion Sickness and Postural Instability. *Ecological Psychology*, 3(3):195–240, 1991.

[63] S. Robertson and T. Bertling. *How To Draw: Drawing And Sketching Objects And Environments From Your Imagination*. Design Studio Press,, 2013.

[64] J. P. Rolland, L. D. Davis, and Y. Baillot. A Survey Of Tracking Technologies For Virtual Environments. In *Fundamentals of wearable computers and augmented reality*, pages 83–128. CRC Press, 2001.

[65] K. Satter and A. Butler. Competitive Usability Analysis Of Immersive Virtual Environments In Engineering Design Review. *Journal Of Computing And Information Science In Engineering*, 15(3), 2015.

[66] J. H. Seo, M. Bruner, and N. Ayres. Aura Garden: Collective and Collaborative Aesthetics of Light Sculpting in Virtual Reality. In *Extended Abstracts of the 2018 CHI Conference on Human Factors in Computing Systems - CHI '18*, pages 1–6, 2018.

[67] C. Shao, A. Bousseau, A. Sheffer, and K. Singh. CrossShade: Shading Concept Sketches Using Cross-Section Curves. *ACM Transactions on Graphics*, 31(4), July 2012.

[68] J. S. Sobel, A. S. Forsberg, D. H. Laidlaw, R. C Zeleznik, D. F. Keefe, I. Pivkin, G. E. Karniadakis, P. Richardson, and S. Swartz. Particle Flurries. *IEEE Computer Graphics and Applications*, 24(2): 76–85, 2004.

[69] I. E. Sutherland. Sketchpad: A Man-Machine Graphical Communication System. *Simulation*, 2(5):R–3, 1964.

[70] I. E. Sutherland. A Head-Mounted Three Dimensional Display. In *Proceedings of the December 9-11, 1968, Fall Joint Computer Conference, Part I*, AFIPS '68 (Fall, part I), page 757–764, 1968.

[71] H. M. Traquair. An Introduction to Clinical Perimetry, Chpt. 1. *London: Henry Kimpton*, pages 4–5, 1938.

[72] Tvori Inc. Shapes XR. https://www.shapesxr.com/, 2021.

[73] Vive. Cosmos Elite headset specs. https://www.vive.com/us/product/vive-cosmos-elite-headset/specs/, 2020.

[74] P. Wacker, A. Wagner, S. Voelker, and J. Borchers. Physical Guides: An Analysis of 3D Sketching Performance on Physical Objects in Augmented Reality. In *Proceedings of the Symposium on Spatial User Interaction*, SUI '18, page 25–35, 2018.

[75] M. Ward, R. Azuma, R. Bennett, S. Gottschalk, and H. Fuchs. A Demonstrated Optical Tracker with Scalable Work Area for Head-Mounted Display Systems. In *Proceedings of the 1992 Symposium on Interactive 3D Graphics*, I3D '92, page 43–52, 1992.

[76] C. Wienrich, C. K. Weidner, C. Schatto, D. Obremski, and J. H. Israel. A Virtual Nose as a Rest-Frame - The Impact on Simulator Sickness and Game Experience. In *2018 10th International Conference on Virtual Worlds and Games for Serious Applications (VS-Games)*, pages 1–8. IEEE, 2018.

[77] E. Wiese, J. H. Israel, A. Meyer, and S. Bongartz. Investigating the Learnability of Immersive Free-Hand Sketching. In *Proceedings of the Seventh Sketch-Based Interfaces and Modeling Symposium*, SBIM '10, page 135–142, 2010.

[78] H. Xia, S. Herscher, K. Perlin, and D. Wigdor. Spacetime: Enabling Fluid Individual and Collaborative Editing in Virtual Reality. In *Proceedings of the 31st Annual ACM Symposium on User Interface Software and Technology*, UIST '18, page 853–866, 2018.

[79] Y. Xia, J. Li, L. Qi, and H. Fan. Loop Closure Detection For Visual SLAM Using PCANet Features. In *2016 International Joint Conference on Neural Networks (IJCNN)*, pages 2274–2281. IEEE, 2016.

[80] B. Xu, W. Chang, A. Sheffer, A. Bousseau, J. McCrae, and K. Singh. True2Form: 3D Curve Networks from 2D Sketches via Selective Regularization. *ACM Transactions on Graphics*, 33(4), July 2014.

[81] A. Young and J. R. Stafford. Real-time User Adaptive Foveated Rendering, U.S. Patent No. 10,192,528. 29 Jan. 2019.

[82] S. Zhai, P. Milgram, and W. Buxton. The Influence of Muscle Groups on Performance of Multiple Degree-of-Freedom Input. In *Proceedings of the SIGCHI Conference on Human Factors in Computing Systems*, pages 308–315, 1996.

7

Input Processing and Geometric Representations for 3D Sketches

Johann Habakuk Israel[1], Mayra Donaji Barrera Machuca[2], Rahul Arora[3], Philipp Wacker[4], and Daniel Keefe[5]

[1]Hochschule für Technik und Wirtschaft Berlin,
University of Applied Sciences, Germany
[2]Dalhousie University, Canada
[3]University of Toronto, Toronto, Canada
[4]RWTH Aachen University, Germany
[5]University of Minnesota, US

Even though some commercially available 3D sketching systems already exist, the need to develop one's own systems may emerge, for example in the context of research projects, in the context of teaching, or to extend the functionality of existing immersive systems with 3D sketching features. After the unique characteristics of 3D sketching were presented in the previous chapter, this chapter outlines the essential steps for generating a 3D sketch using freehand user input. Many of the principles of 2D sketching can also be applied to 3D and will not be repeated here. Instead, selected methods that have proven themselves in practice in 3D are explained.

As immersive 3D sketching systems are highly interactive, all steps must be performed in real time and post processing approaches are not suitable here. Besides the hardware requirements described in Section 6.1, the following steps must be taken into account on the software side when realizing a freehand 3D sketching system:

tracking: recording the position, orientation, and states of the buttons of the interaction devices used (see Section 6.1.2),

filtering: filtering of the input data, for example, smoothing,

sampling: transfer of the input data into an internal data structure, if necessary removal of unnecessary supporting points,

mesh creation: creation of a representable mesh based on the supporting points, for example by extrusion of a basic shape (brush), and

rendering: real-time representation of the mesh

Immersive sketching systems typically provide additional features such as deleting, editing, saving, and loading sketch entities. Since these are not significantly different from 2D sketching, they are not discussed here. In addition, immersive sketching systems typically have 3D menus for selecting and parameterizing specific sketching functions such as brush, color, and curve properties, among others. An overview of the design of 3D menus is provided by Dachselt and Hübner [1] and LaViola et al. [2]. Further user interface-related guidance on processing user input to control immersive 3D sketching systems can be found in Jackson and Keefe [3].

7.1 Filtering

As mentioned in Section 6.1.2, the input devices used in 3D sketching usually provide data in the full six spatial degrees of freedom. Many methods for generating stroke geometry take the orientation into account, for example, in the creation of calligraphic sketches. Here the orientation of the interaction device has an important influence on the resulting geometry. The filter methods described here only refer to translational data. Filter methods also exist for rotational data but are much more sophisticated. Hartley et al. [4] provide a good overview of corresponding methods. Some 3D frameworks also offer the possibility to interpolate rotations using quaternions.

In freehand 3D sketching, input data is typically filtered for two reasons, namely to reduce or smooth out hardware-induced noise, typically caused by the inaccuracy of the tracking systems used, and to reduce small jittery movements caused by the user when guiding the 3D input devices.

Smoothed user input produces much more aesthetic and satisfying sketches than unfiltered ones. At the same time, filtering causes slight delays, which add to the delays already caused by the tracking and rendering system. This is especially noticeable with fast sketch movements, such as when drawing wavy lines quickly. In AR and projection-based environments such as CAVEs and powerwalls, users typically see the physical input devices directly. Here, the delay is visible in such a way that during sketch movements, the extrusion point of the virtual ink follows the position of the physical input device or the physical pen tip at a noticeable distance. In HMD-based systems, where users hold the interaction devices in their hands but typically only see them as rendered 3D geometries in the virtual environment, the pen tip and extrusion point are always synchronous. On the other hand, there can be a disturbing offset between the position of the virtual input device and the position of the physical input device held in the hand, which users perceive although they cannot see the physical input device directly, but feel where and how they are holding it [visual-proprioceptive conflict, cf. 5].

There are several filtering methods that differ in particular in the degree of smoothing and delay of user input[1]. The easiest way to filter user input is the Moving Average Filter. This filter uses a certain number of the last positions X_n measured by the interaction device and forms a simple average value $\bar{X}_n$. Although good smoothing can be achieved with this filter, the delay increases noticeably with the number of positions considered. In addition, sudden changes in direction of the pen guidance cannot be satisfactorily represented by this filter. To address this problem, the Exponential Moving Average Filter, given by Equation 7.1, can be used. This filter calculates the filtered value $\bar{X}_n$ using a weighted average of the current position X_n of the interaction device and the last filtered value $\bar{X}_{n-1}$. The higher the weighting factor a weights the current position, the higher the responsiveness to motion changes, but with a lower smoothing effect.

$$\bar{X}_n = aX_n + (1-a)\bar{X}_{n-1} \tag{7.1}$$

This filter reacts better to changes in direction, but still generates noticeable delays when smoothing is strong. Double Exponential Smoothing Filters, defined by Equation 7.2 and Equation 7.3, offer an even better response to rapidly changing directions of motion [6]. They include the trend (b_n) of the movement, which is determined by an average of the difference between the last two filter results ($\bar{X}_n$, $\bar{X}_{n-1}$) and the last calculated trend, weighted with the factor γ.

$$b_n = \gamma(\bar{X}_n - \bar{X}_{n-1}) + (1-\gamma)b_{n-1} \tag{7.2}$$
$$\bar{X}_n = aX_n + (1-a)(\bar{X}_{n-1} + b_{n-1}) \tag{7.3}$$

γ factors closer to 1 are appropriate for reliable tracking systems with less noise and result in shorter delays. For less accurate tracking systems, such

---

[1]For immediate feedback it is possible to use a temporary point (typically X_n) and replace it with the filtered point ($\bar{X}_n$) when the filtering is done.

as magnetic tracking systems, γ factors lower than 0.1 may be necessary to achieve smooth lines, resulting in noticeable spatial lag. The delay becomes less noticeable if the tracking system provides a high updated rate, that is, many samples per second. Reading the tracking system with the highest possible frequency is, therefore, especially helpful if the tracking data is noisy. Usually, this requires a process that is independent of the rendering thread, which runs at a higher frequency and writes the tracking data into a queue. As mentioned in Section 6.2.2.2, research by Keefe et al. [7], Arora et al. [8], Barrera Machuca and Stuerzlinger [9], and Batmaz et al. [10] indicates that freehand sketches are often noisier in depth than in horizontal and vertical directions. Therefore, it may be beneficial to smooth tracking data more strongly in the egocentric depth direction.

Another useful method to improve the visual quality of sketches was proposed by Thiel et al. [11]. To distinguish details intended by the drawer from noise caused by the tracking system, they use the speed at which the line was drawn. The higher the speed of the drawing movement, the more the user input is smoothed.

7.2 Sampling

As soon as the user presses the button provided on the interaction device, the position and orientation data supplied by the tracking system are recorded to create the sketch geometry. Typically, this data is filtered as described above before it is processed further. The next step in the processing chain is called sampling. This is where it is decided which of the input data will be used to create the geometry. It is important not to use all of the data generated by the tracking system, otherwise, a lot of data could be generated that has no visual effect. For example, if the user holds the input device still with the button pressed, approximately identical position data is generated in each scanning run. If all of this data were to be transferred, the result would be many lines in a very small space that would hardly be recognizable to the observer. In addition, a large number of points in a single spot could distort the appearance of the line in a way not desired by the user. At the same time, the data set to be processed (supporting points of the sketch) would be considerably inflated, so that all subsequent processes would be unnecessarily burdened.

In the sampling process, therefore, only those points are considered that are sufficiently far apart to be perceived by the user. The easiest way to do this is to define a minimum distance that points must maintain from each other. For each position supplied by the tracking system, the Euclidean distance to the previous position is then determined. The new point will only be considered if this distance is greater than the minimum distance. Furthermore, it is possible to check whether several consecutive points of a line lie on a straight line. All points to which this applies can be removed. An efficient method for selecting the relevant position data for generating strokes is the Douglas-Peucker algorithm [12] (refer to Chapter 3 for a brief description of this algorithm).

In some cases, the Nyquist frequency [13] may be exceeded, which should be at least twice the frequency of the signal. In other words, the system does not collect enough position data to correctly capture and accurately reproduce the user's sketching movements. For example, when the user performs very fast zigzag movements, and the created path does not accurately reproduce the user's sketch movements. In this case, too little sampling is often not due to too slow tracking technology but results from an unfavorably implemented processing pipeline. To achieve better results, developers can delegate the retrieval of tracking data to a parallel process.

7.3 Geometric Representations

After the user's pen or sketch movements are captured and filtered, they must be stored in an internal data structure of the sketching application. Many 3D frameworks offer specific functionalities for this purpose, for example, in Unity3D a so-called line renderer exists, which creates a geometric representation from a list of points and renders it by connecting the points with straight lines [14]. The disadvantage of using such functionalities is that many dependencies to the used 3D framework arise in the program code and adjustments are more difficult to implement.

The simplest way to do this oneself is to use the scene graph of the 3D framework used. Incoming motion information is then converted directly into representable 3D geometry, typically by extrusion (see below). Since the scene graph typically only stores the geometric representations, but not the original sketch movements, this variant has a clear disadvantage that changes to sketches and the loading and saving of sketches are very difficult. It is, therefore, a good idea to save the (filtered) sketch motion data and use it to generate renderable sketch elements in a separate step. Since in this case, the saved sketch entities are the parameters of the generated geometry, this approach can be described as parametric modelling [15]. Parametric modelling is already a standard procedure in CAD. Every CAD tool has a so-called modelling kernel, which realizes the processing and manipulation of 3D model data as well as the generation of displayable 3D elements, a process which is called tessellation. In the area of open source software, the modelling core Open Cascade is available [16].

For smaller projects, which only need a few sketch element types, such as freehand strokes, straight lines or simple Bézier surfaces, it is also possible to develop one's own modelling kernel which should have an interface for generating sketch elements, for manipulating them and for persistence (loading and saving).

Section 8.2.1.1 describes a variety of basic geometric shapes that are possible and conceivable in immersive sketching applications. This section focuses on the generation of strokes, the most basic geometric elements of sketches. In addition, other elements such as surfaces or volumetric objects can be created in a sketching manner. A number of excellent textbooks are available for more detailed illustrations of internal data structures, parameterization, mesh creation, and rendering [e.g. 17, 18, 19].

7.3.1 Strokes

Strokes are the most important geometric elements of sketches. There are several ways to generate internal data representations from the information about sketch movements and display them. The simplest possibility is to save the support points generated in the sampling process for each line in a list and to connect them to each other with a cylinder or a straight tube. If the support points are close together, this method can already provide satisfactory results. However, in the case of fast sketch movements which, due to the limited sampling rate of the tracking system, provide support points that are comparatively far apart, visible corners and edgy curve shapes can occur in this method. To avoid this problem, the use of parametric curves is recommended. These describe the shape of the curve using a mathematical function that receives the control points as input values. Along this curve a basic shape is then extruded so that renderable elements are generated; a process called tessellation.

7.3.1.1 Parametric Curves

If the support points of a line resulting from the sampling process are described by $s_0, \cdots, s_n$, where n is the number of support points, then a parametric curve can be created using a function $p(t)$, which receives a subset of the support points as input and calculates the points $p_0, \cdots, p_m$ of a line. The number of support points n and the number of line points m typically differ. This makes it possible to draw lines with higher resolution than the interpolation points would actually provide. However, this is accompanied by the fact that the drawn line can deviate from the real line, since the lines must be synchronous only at the support points. A selection of simple parametric curves is presented below. A comprehensive overview can be found, for example, in Akenine-Möller et al. [19, chapter 17] .

7.3.1.2 Linear Interpolation

The simplest and still, at least in research projects, widely used form of a parametric curve is a linear interpolation between two interpolation points. For this purpose, the slope between the two points is calculated. Each point of the curve can then be determined using Equation 7.4.

$$p(t) = s_{i-1} + t(s_i - s_{i-1}), \ \ t \in [0, 1]. \tag{7.4}$$

The result of the linear interpolation is a piece-wise linear curve, which connects all points by lines. Since no new line points p can be determined by the linear interpolation, which does not already lie on a straight line between $s_i - s_{i-1}$, a parametric description is usually omitted here.

7.3.1.3 Bézier Curves

The jagged stroke pattern achieved by linear interpolation is not sufficient in many application contexts. Smooth, aesthetic transitions between the line segments, or, in mathematical terms, a geometric continuity of order G^1 (tangent continuity) or even G^2 (curvature continuity), are often expected. To achieve this, Bézier curves which give G^1 continuity are used in many cases [20] (see Box 2.6 in Section 2.2 for further details on Bézier curves). These curves also use the data from the sampling process as support points. According to the degree k of the Bézier curve, $k + 1$ points are used to calculate a curve segment. The resulting curve lies within a convex hull of the control points and runs through the first and last of the control points. The continuous progression is achieved by weighting the influence of the points S_i on the resulting curve differently using the Bernstein polynomial B_i^k according to the parameter t. Equation 7.5 gives an example of a cubic Bézier curve drawn between four points. In computer graphics, cubic Bézier curves with $k = 3$ are most frequently used.

$$p(t) = B_0^3(t)p_0 + B_1^3(t)p_1 + B_2^3(t)p_2 + B_3^3(t)p_4, \ \ t \in [0, 1]. \tag{7.5}$$

In Bézier curves, the resulting curve is most strongly pulled towards the support points to which it is closest, but without reaching them. While this results in smoothly continuous curves, it is also the biggest problem with using Bézier curves for sketches since the curve shown does not pass through the curve described by the stylus, except at the first and last point of each curve segment, but only approximates it. This deviation can result in the visual appearance of the sketch not corresponding to the user's intention. The fact that some of the support points are not on the stroke can also be problematic if users want to change their sketch later. For example, if users wanted to change the curve, the easiest way to do this would be to move the support points accordingly. However, if they do not lie directly on the sketched line, these manipulations become unnecessarily difficult. For this reason, representations should be used for the visualization of sketched lines, where the support points lie directly on the sketched line, which can be achieved by Cubic Hermite interpolation.

7.3.1.4 Cubic Hermite Interpolation

The advantage of Cubic Hermite interpolation[2] is that it is relatively easy to control [19]. It is sufficient to define a tangent m_k at each support point p_k. The sketched line can easily be transformed into individual Hermite curve segments. Each support point is then both the end and the beginning of a segment. The tangent of each interpolation point can be calculated by the vectors from the previous p_{k-1} to the current interpolation point p_k and from the current interpolation point p_k to the following interpolation point p_{k+1} as defined in Equation 7.6.

$$m_k = \frac{p_k - p_{k-1}}{2} + \frac{p_{k+1} - p_k}{2} \tag{7.6}$$

The tangents can then be used to interpolate the sketched line using the cubic polynomials in Equation 7.7.

$$p(t) = (2t^3 - 3t^2 + 1)p_0 + (-2t^3 + 3t^2)p_1 + (t^3 - 2t^3 + t)m_0 + (t^3 - t^2)m_1, \tag{7.7}$$

where $t \in [0, 1]$. Figure 7.1 shows an example curve where the support points have been highlighted.

---

[2]Cubic Bezier and Cubic Hermite interpolation are mathematically equivalent and can be translated into each other.

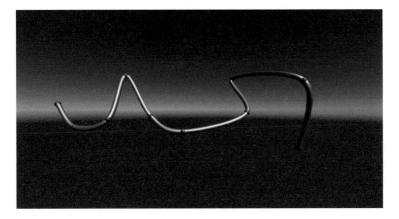

Figure 7.1 Piecewise Hermite curve with 10 control points and 9 segments

7.3.2 Extrusion

So far we have only discussed how to determine the support points of a line and the points in between. However, this is not enough for the representation in a 3D environment. Here, lines must be created that have a volume, even if it is very small. Typically, tube-shaped geometries are created during 3D sketching, but there are also other possibilities, which will be discussed in the following.

The basic idea of extrusion is to create a three-dimensional body in which a two-dimensional basic shape is drawn along a path through space. An analogy in the real world is for example the creation of soap bubbles. Here, the tube soaked in lye is the basic form from which the (albeit slightly dented) soap bubble is extruded as soon as the tube is moved or air blows through it. Theoretically, basic shapes can be designed arbitrarily, but should not intersect themselves. They can either be defined in advance and read in via configuration files or drawn interactively by the user so that they can configure their own drawing tools. Extrusion is a basic modelling technique of CAD. While extrusion paths are precisely parametrically described in CAD, 3D sketching uses freehand pencil movements for extrusion. The basic shape required for extrusion is typically described by a list of two-dimensional points as shown in Figure 7.2(a). These are projected into three-dimensional space at each support point of the line in such a way that the respective support point and the center of the basic shape are superimposed. In a further step, the points created are linked with the points

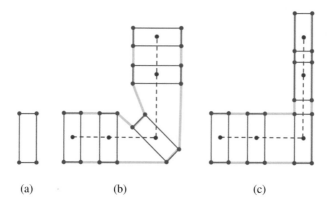

Figure 7.2   Extrusion: (a) basic shape, (b) extrusion perpendicular to the tangent of the sketched line, (c) calligraphic extrusion

of the last extrusion step to create displayable 3D elements (see tessellation below). There are several approaches to the question of how the basic shape should be oriented during extrusion.

7.3.2.1 Extrusion While Maintaining the Diameter of the Basic Shape

If the diameter of the basic shape is to be retained as far as possible, it must be extruded perpendicular to the tangent of the sketched line. In the case of very sharp lines, the extruded geometry may intersect itself and therefore a uniform diameter cannot be guaranteed, however, this method usually gives satisfactory results as shown in Figure 7.2(b).

7.3.2.2 Calligraphic Extrusion

For other applications, it may be necessary to transfer the 3D orientation of the stylus one-to-one to the extruded geometry. Similar to a calligraphic pen, different geometric shapes can be extruded depending on the direction of movement. This is typically achieved by transforming the two-dimensional basic shape in each extrusion steps into the respective local coordinate system of the interaction device as shown in Figure 7.2(c).

7.4 Processing of Further User Input

As mentioned in Section 8.2.2.1, besides the position and orientation of the drawing tools, other input parameters can be used to influence the shape of a sketch. For example, it is possible to use a force sensor to measure how hard the user presses the pen and to determine the width of the drawn line by scaling the basic shape on its y-axis [cf. 21]. There are no limits to creativity when mapping the input parameters to the created shape. It is possible, for example, to arrange pressure sensors radially on the pen and to manipulate the extruded basic shape according to the pressure measurements so that the user has even more control over the created shape while drawing. The speed of the pen movements can be used as a further input parameter for geometry creation. For example, fast movements can create narrow and slow movements in thick shapes or, as with brush techniques, denser or looser shapes.

7.5 Tessellation and Rendering

Parametric curves can have infinite resolution. In the tessellation process, polygons (usually triangles) are created from this continuous description in order to output them to the display via the 3D graphics pipeline. The smaller the step size in the tessellation process, the finer the resolution of the 3D geometry, but also the more triangles are created, which can result in performance losses.

The tessellation process can be performed on the GPU (available through DirectX 11 and OpenGL 4.0) or on the central processing unit (CPU). Although the first method is faster, the second method is often chosen for compatibility reasons.

When generating line geometries, the tessellation and extrusion processes are often closely related. First, in each extrusion step, a point of the basic shape is transformed into 3D space as described above. Then the points of the basic shape (vertices) of the current extrusion step and the previous extrusion step have to be linked in such a way that representable geometries are created. For this purpose, a mesh is created that consists of 3D points (vertices), connections (edges), and surfaces (usually triangles). Triangles are defined by specifying the indices of the points to be connected.

If a sketch basic shape with n vertices is given and $v(i, 0 \ldots n - 1)$ describe the vertex indices of the current basic shape and $v(i - 1, 0 \ldots n - 1)$ the vertex indices of the basic shape of the previous extrusion step, the required triangles can be created in a loop:

```
// create triangles between all vertex-pairs except
// between the first and the last
```

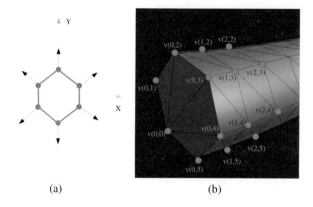

(a)   (b)

Figure 7.3   3D extrusion: (a) basic shape consisting of six points and (b) extrusion of the basic shape.

```
for (a = 0; a < n-2; a++) {
    create_triangle( v(i, a), v(i-1, a), v(i, a+1));
    create_triangle( v(i, a+1), v(i-1, a), v(i-1, a+1));
}

// close the band: create triangles between the first
// and the last vertex pairs
create_triangle( v(i, n-1), v(i-1, n-1), v(i, 0));
create_triangle( v(i, 0), v(i-1, n-1), v(i-1, 0));
```

The result could look like Figure 7.3. If one sets $n = 2$, the for loop is ignored and a flat ribbon is generated. This is useful if one wants to draw lines without thickness. Note that the controller orientation can still be used to align the ribbon.

The points of the triangles can be defined clockwise or counterclockwise. This defines which of the sides of the triangle point outwards and which point inwards. In OpenGL, by default, the counterclockwise sides of the triangle point outward.

For the first (when the user presses the draw button) and last extrusion step (when the user releases the button), a cover should also be drawn on the beginning and end of the stroke so that it appears as a closed body as shown by the four triangles between the points $v(0, 0) \ldots v(0, 5)$ in Figure 7.3.

It is also important to set a surface normal for each vertex. Otherwise, the sketched lines will be displayed in one color and lose all shading.

The procedure described here is particularly easy and quick to implement, but can produce unattractive results in the case of unfavorable curvatures and self-penetration. More advanced subdivision techniques that adapt the shape of the triangles to the actual shape of the curve at higher resolutions are, for example, Adaptive Tesselation and Fast Catmull-Clark Tesselation. These are explained in detail for example in Akenine-Möller et al. [19].

7.6 Modelling Kernels

As an alternative to the methods presented here, some systems [e.g. 22, 23] outsource the internal geometry presentation as well as the tessellation process to external modelling kernels. Open source modelling kernels such as OpenCascade [16] is often used for this purpose. These offer the possibility to parametrically describe complex geometric bodies and to tessellate them with arbitrary levels of detail. This can be used, for example, to provide objects with different levels of detail depending on the distance to the viewer. In addition, they also support the import of CAD files, which allows them to be displayed within the 3D scene. This point is particularly advantageous when the sketching functionality is to be integrated into a CAD infrastructure.

A disadvantage of external modelling kernels is that they often require high integration and maintenance efforts. The decision to develop one's own lightweight modelling kernel or to integrate an external modelling kernel should, therefore, be made taking into account the available development capacities, the requirements for interoperability and the expected complexity of the sketches produced.

7.7 Summary

In this chapter, methods for processing freehand user interactions and techniques for transforming them into 3D sketches were described. However, only basic principles could be explained. Further issues, such as texturing or editing of 3D sketches were not mentioned. For more detailed descriptions, interested readers are referred to related literature from the field of computer graphics [e.g. 17, 18, 19].

The principles presented in this chapter can help to better understand the functionalities of the 3D sketching applications presented later in Chapter 9. Prior to this, Chapter 8 explains which interaction techniques and devices are available to users to create 3D sketches.

References

[1] R. Dachselt and A. Hübner. A Survey And Taxonomy Of 3D Menu Techniques. In *Proceedings of the 12th Eurographics Conference on Virtual Environments*, EGVE'06, page 89–99, 2006. Eurographics Association. ISBN 978-3-905673-33-3.

[2] J. J. LaViola, E. Kruijff, R. P. McMahan, D. Bowman, and I. P. Poupyrev. *3D User Interfaces: Theory and Practice*. Addison-Wesley, Boston, 2 edition, 4 2017.

[3] B. Jackson and Daniel F. Keefe. From Painting to Widgets, 6-DOF and Bimanual Input Beyond Pointing. In *VR Developer Gems*. A K Peters/CRC Press, Boca Raton, FL, USA, 2019.

[4] R. Hartley, J. Trumpf, Y. Dai, and H. Li. Rotation Averaging. *International Journal on Comput Vision*, 103:267–305, January 2013.

[5] Y. Lee, I. Jang, and D. Lee. Enlarging Just Noticeable Differences Of Visual-proprioceptive Conflict In VR Using Haptic Feedback. In *2015 IEEE World Haptics Conference (WHC)*, pages 19–24. 2015 IEEE World Haptics Conference (WHC), 6 2015.

[6] NIST/SEMATEC. Double exponential smoothing, 2012. URL https://www.itl.nist.gov/div898/handbook/pmc/section4/pmc433.htm. [Online; accessed 2020-03-09].

[7] D. Keefe, R. Zeleznik, and D. Laidlaw. Drawing on Air: Input Techniques for Controlled 3D line Illustration. *IEEE Transactions on Visualization and Computer Graphics*, 13(5):1067–1081, 2007.

[8] R. Arora, R. H. Kazi, F. Anderson, T. Grossman, K. Singh, and G. Fitzmaurice. Experimental Evaluation of Sketching on Surfaces in VR. In *Proceedings of the 2017 CHI Conference on Human Factors in Computing Systems*, CHI '17, page 5643–5654, 2017.

[9] M. D. Barrera Machuca and W. Stuerzlinger. The Effect of Stereo Display Deficiencies on Virtual Hand Pointing. In *Proceedings of the 2019 CHI Conference on Human Factors in Computing Systems*, CHI '19, 2019.

[10] A. U. Batmaz, M. D. Barrera Machuca, D. M. Pham, and W. Stuerzlinger. Do Head-Mounted Display Stereo Deficiencies Affect 3D Pointing Tasks in AR and VR? In *2019 IEEE Conference on Virtual Reality and 3D User Interfaces (VR)*, pages 585–592. IEEE, 2019.

[11] Y. Thiel, K. Singh, and R. Balakrishnan. Elasticurves: Exploiting Stroke Dynamics and Inertia for the Real-time Neatening of Sketched 2D

Curves. In *Proceedings of the 24th Annual ACM Symposium on User Interface Software and Technology*, page 383–392, 2011.

[12] D. H. Douglas and T. K. Peucker. Algorithms for the Reduction of the Number of Points Required to Represent a Digitized Line or its Caricature. In *Classics in Cartography: Reflections on Influential Articles from Cartographica*, page 15–28, 2011.

[13] C.E. Shannon. Communication In The Presence Of Noise. *Proceedings of the IEEE*, 86(2):447–457, February 1998. ISSN 1558-2256.

[14] Unity. Scripting API: LineRenderer, 2020. URL https://docs.unity3d.com/ScriptReference/LineRenderer.html.

[15] F. Fu. Design and Analysis of Complex Structures. In Feng Fu, editor, *Design and Analysis of Tall and Complex Structures*, pages 177–211, 1 2018.

[16] OpenCascade. Open CASCADE technology, the open source 3D modeling libraries | collaborative development portal, 2018. URL https://dev.opencascade.org/. [Online; accessed 2018-01-24].

[17] J. D. Foley, A. Dam, and S. K. Feiner. *Computer Graphics: Principles and Practice*. Addison-Wesley, Upper Saddle River, New Jersey, 3 edition, July 2013. ISBN 978-0-321-39952-6.

[18] M. Botsch, L. Kobbelt, M. Pauly, P. Alliez, and B. Levy. *Polygon Mesh Processing*. A K Peters, Natick, Mass, October 2010. ISBN 978-1-56881-426-1.

[19] T. Akenine-Möller, E. Haines, N. Hoffman, A. Pesce, M. Iwanicki, and S. Hillaire. *Real-time Rendering*. Taylor & Francis, CRC Press, Boca Raton, fourth edition edition, 2018. ISBN 978-1-138-62700-0.

[20] Eric W. Weisstein. Bézier curve, 2020. URL https://mathworld.wolfram.com/BezierCurve.html. source: mathworld.wolfram.com publisher: Wolfram Research, Inc.

[21] J. H. Israel, E. Wiese, M. Mateescu, C. Zöllner, and R. Stark. Investigating Three-dimensional Sketching For Early Conceptual Design—results From Expert Discussions And User Studies. *Computers & Graphics*, 33(4):462 – 473, 2009.

[22] P. Fehling, F. Hermuth, J. H. Israel, and T. Jung. Towards Collaborative Sketching in Distributed Virtual Environments. In *Kultur und Informatik*, page 253–264, 2018.

[23] VENTUS. Virtual Environment for Teamwork and Ad-hoc Collaboration Between Companies and Heterogeneous User Groups, 2017. URL http://ventus3d.com/. [Online; accessed 2017-11-09].

8

Interaction Devices and Techniques for 3D Sketching

Mayra Donaji Barrera Machuca[1], Rahul Arora[2], Philipp Wacker[3], Daniel Keefe[4], and Johann Habakuk Israel[5]

[1]Dalhousie University, Canada
[2]University of Toronto, Toronto, Canada
[3]RWTH Aachen University, Germany
[4]University of Minnesota, US
[5]Hochschule für Technik und Wirtschaft Berlin,
University of Applied Sciences, Germany

In the previous chapter, we discussed the different ways to create a stroke. In this chapter, we present various interaction devices and techniques that help users sketch in a 3D environment. An interaction device or technique is a combination of hardware and software elements that help users accomplish a single task [1], in this case, to create 3D sketches. When talking about specific interaction devices or techniques, it is important to consider the needs of the user. For 3D sketching, these needs change depending on the type of sketch the user wants to create [2, 3, 4]. It is also important to consider the affordances that different interaction devices or techniques have, as they affect the way users utilize them [5, 6]. Affordances are an attribute of the tool that shows which features they offer to the user [5, 6].

Based on these considerations, we divide this chapter into two parts. First, we will talk about the user interface or interaction techniques that allows users to draw conceptual and technical sketches. We will discuss the characteristics that make them more appropriate for each specific sketch type. Then, we will talk about the functional and physical affordances of a user interface or interaction technique. Here, we will discuss how these affordances can affect the sketch produced by them. This chapter is useful for readers that want to design their interaction device or technique for sketching in 3D, as we discuss

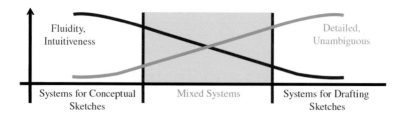

Figure 8.1   The three different types of systems we are discussing in this chapter.

the different properties they need to consider. For example, the type of sketch their interaction device or technique can create and the affordances they need to have.

8.1 Sketch Types

In Chapter 1, we already talked about the design process, which is "the complex activity of creating and evaluating specifications of artifacts whose form and function achieve stated objectives and satisfy specified constraints" [7] and the importance of sketching during it. In general, during the design process, the goal of the user for the sketches they make changes depending on their current stage, and this goal informs the sketch type [4]. For example, a designer produces a conceptual sketch when exploring a new idea, and a technical drawing when they are creating the final design of a product. Here, we divide the interaction devices and techniques into three groups (Figure 8.1) based on their functionalities:

(a) systems for conceptual sketches, which are unstructured and ambiguous,
(b) systems for drafting that produce accurate scale drawings,
(c) mixed systems that help users draw sketches with characteristics of the other two sketch types

The systems in this last category are those that see 3D sketching as a unique medium and try to create a balance between keeping the interaction fluid and making accurate sketches.

8.1.1 Conceptual 3D Sketches

In Chapter 1, we state that conceptual sketches are used early in the design process to communicate the general aesthetic and to explore the technical

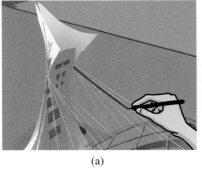
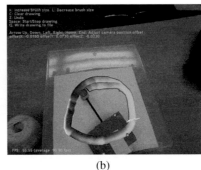

(a)   (b)

Figure 8.2   3D freehand drawing systems follow the user hand movements. For example, (a) SymbiosisSketch [11] and (b) Yee et al's [12] system.

aspects of a design. In this stage, designers need to explore initial ideas quickly and simply, so they can formulate a tentative solution. Designers of 3D sketching interfaces for conceptual sketches need to provide interaction techniques that are fluid, expressive, and fast [8]. Interaction techniques in this category also need to be intuitive [9] and support the creative process, because other windows, icons, menus, pointer (WIMP)-based forms of interaction would fragment and hinder creative designing [10]. One interaction technique that meets these characteristics is freehand drawing, in which the stroke follows the user hand movement. There are different variations of freehand drawing, which depend on the input device used to create the strokes. Each of these implementations provides users with unique ways to draw more accurately.

3D freehand drawing

The first interaction technique we present is a 3D freehand drawing which is the direct translation of sketching with a pen on paper to 3D as illustrated in Figure 8.2. 3D freehand drawing uses a 6-DoF input device to follow the user's arm movements in space, which provides an intuitive and effective method of conceptualizing new shapes [13]. This feature makes this interaction a popular choice among user interfaces [11, 14, 15, 16, 17, 18, 19, 20, 21, 12, 22, 23]. One characteristic of 3D freehand drawings is that the system does not constrain the user's actions. However, a user interface or device can use the stroke's properties to affect the sketch. To name three examples, in Fluid Sketching [17] users paint with fluids that

(a)   (b)

Figure 8.3   SymbiosisSketch [11] allows decreasing (a) or increasing (b) the object's scale to allow users to draw large-scale objects and fine details, respectively.

blend, in CavePainting different brushes have different behaviors [19] while Griddraw [20] creates a 3D grid of force vectors to alter the appearance of the stroke.

A limitation of 3D freehand drawing is the difficulty to control the stroke using the shoulder or the elbow [24, 25]. One way to solve this problem is to allow users to create small strokes that users can draw by only moving their wrist, and then provide tools to scale these strokes to their real size. Such an approach is adopted in SymbiosisSketch [11] and Tano et al.'s system as illustrated in Figure 8.3. However, these scaling techniques should not break the interaction flow by introducing WIMP-like interaction patterns to, for example, manually position the drawing plane. Another way to provide more stroke control is to allow users to draw on a tracked 2D plane inside a 3D virtual environment [11, 15, 26]. These systems benefit from having a physical surface to draw on and latch onto the user's experience of sketching with a pen. Similar to these interaction devices, other systems use a 2D screen to display the sketch [27, 28, 29]. However, these systems need to provide navigation tools so users can change their viewpoints. These systems also need to project the user 2D stroke to 3D. For example, Agile 3D Sketching [27] and SketchingWithHands [28] project the strokes to a virtual plane that is defined by the user. Both systems allow users to move their viewpoint using a combination of gestures.

3D modelling

There is also a category of user interfaces that use interaction devices and techniques to mimic the actions of modelling an object with clay [30, 31, 32, 33, 34]. See Figure 8.4. These tools are more related to modelling than

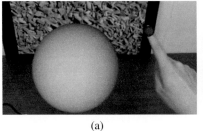 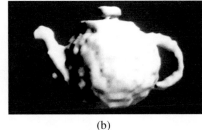

(a)   (b)

Figure 8.4 3D modelling systems emulate the action of modelling with clay. For example, (a) AiRSculpt [33] and (b) Sculpting [31].

drawing, but they still focus on being fluid and expressive by not focusing on the creation of polygons. For example, AiRSculpt [33] allows users to model a sphere into different shapes by adding and removing material. Another example is Leal et al.'s system that provides users with tracked fabric that users can twist and reshape to create virtual surfaces.

In general, previously proposed interaction devices and techniques for 3D sketching conceptual sketches try to stay flexible and fast by constraining the user action in the least amount possible. Designers of interaction devices or techniques for 3D sketching should think of ways to use the advantages of using a digital system, like Fluid Sketching [17] and Griddraw [20]. Also, finding novel ways to merge physical objects and virtual objects [32, 35] is still an open area of research.

8.1.2 Technical 3D Sketches

Drafting sketches, also known as technical sketches, are mostly used during the later stages of the design process to show the final design of a product. In this stage designers need to precisely communicate an idea, so the sketch should be unambiguous and easy to understand. To achieve this, most designers use conventions and standards to represent objects, like the rules for orthographic projection. They also use CAD systems like AutoCAD [36] or Solidworks [37], as these allow to automate the process and bring a high degree of accuracy to the sketch but with the loss of fluid interaction. However, the control of the user interface requires much more attention from the designers than in sketch-based systems. Designers of 3D sketching interactions for drafting sketches need to provide tools that allow users to create precise shapes like lines, circles, and curves. User interfaces in this

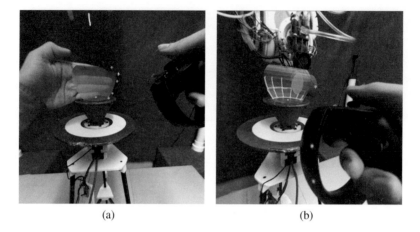

(a)   (b)

Figure 8.5   RoMA [38] is a two-modes system because it allows users (a) to use 3D freehand sketching and (b) to edit the vertices of a shape.

category also need to provide tools to edit the sketch. However, it is important to avoid creating a WIMP-based 3D modelling software and to avoid losing the intuitiveness and fluidity of sketching in 3D by constantly clicking on buttons or menu items.

Bimodal interfaces

One way to fulfill the requirements of drafting sketches is to create a user interface that has two modes, one uses 3D freehand drawing and the other allows users to create geometrical figures, like cubes, cylinders, and spheres [39, 38] or use tools like revolve [40]. These two modes should be separated because the characteristics of each interaction are different. For freehand sketching, the ability to guide the interaction device should be intuitive, which does not require a conscious engagement. However, for creating specific geometric figures, the user needs to consider what each figure represents and its properties [41]. One example of such interfaces is RoMA [38], illustrated in Figure 8.5 which utilizes 3D freehand drawing as the main interaction technique. The system also allows users to create traditional CAD primitives using revolve, extrude, loft, and sweep operations. By mixing both interaction techniques, the final product is accurate enough that a robotic 3D printer can fabricate the objects drawn by the user. Another example is Holosketch [39] which allows users to create strokes using

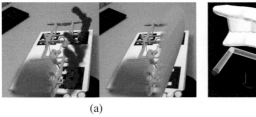
(a)

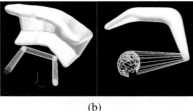
(b)

Figure 8.6 Automatic creation of 3D meshes systems use algorithms to translate the user hand movement in space into 3D meshes. For example, (a) Fuge et al.'s system and (b) Brush2Model [45].

freehand 3D drawing. Users can also create primitives like cylinders, spheres, and cubes. By mixing both interaction techniques, users can create complex shapes without having to sketch every single object. In summary, these interfaces have one mode where users can sketch without constraints and another mode where users can reflect on what was drawn and refine it using widget-based interactions like handles and sliders.

Automatic creation of 3D meshes

Another way to create drafting sketches is to use an interaction technique that translates user movement in space into 3D meshes. In these interaction techniques, users create strokes using 3D freehand drawing. Then, the system translates these strokes into meshes [40, 42, 43]. Most systems aim to create the new surface without delay, but usually, the time between the stroke and the mesh creation depends on the computational power needed to calculate the new mesh and on the properties of the hardware used to run the system. One problem with this approach is that a noticeable delay between the two might affect the interaction by losing the fluidity of 3D freehand sketching. Other interaction techniques allow users to define the position and shape of a surface in space. For example, Fuge et al.'s system illustrated in Figure 8.6(a) automatically converts a cloud of points in space into surfaces. Users create these clouds by tapping with their fingers in the air using a custom-made glove to detect the hand gestures and the pressure of each tapping. Brush2Model [45], shown in Figure 8.6(b) gives users the ability to draw the skeleton of a 3D object. The system automatically covers this skeleton with a surface.

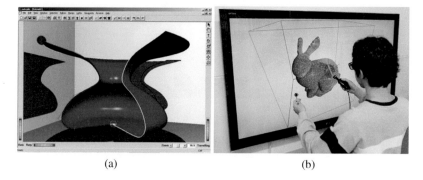

(a)   (b)

Figure 8.7   Systems for drafting sketches should allow users to edit their strokes. For example, (a) by directly editing the curve [46], and (b) by over-sketching the existing mesh [42].

Editing sketches

User interfaces for drafting sketches also allow users to edit their sketches. A common approach is to provide the standard tools used in modelling software, like tools to manipulate the vertices of the object [44, 42, 45], change the scale, position, and orientation of the object [39], or manually modify the curves [46] as illustrated in Figure 8.7(a). Another approach is to allow users to over-sketch the existing meshes [46, 42, 43], with this technique the user re-draws a new line over the previous one to correct any mistakes as illustrated in Figure 8.7(b). Giving users the ability to over-sketch their designs, allows them to fix their mistakes without breaking the sketch-like interaction, as over-sketching is widely used as a design tool for pen and paper drawing [47].

Designers of interaction devices or techniques for 3D sketching should consider different ways to use the advantages of sketching in 3D while simultaneously giving users tools to edit their strokes in-depth. This way, users can fix any error they made, like in RoMA [38] and SPACESKETCH [42]. Future interaction devices or techniques should also try to create systems that allow users to focus on the design process and less on how to create the surface [44, 45].

8.1.3 Mixed Interfaces

Mixed interfaces aim to enable users to accurately sketch objects without losing the advantages of sketching in a 3D space, such as fluidity, expressiveness, and speed. These interfaces limit the user actions to make

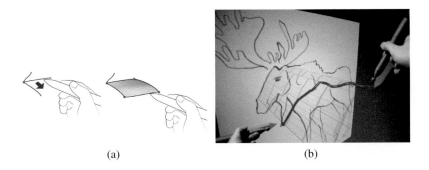

(a)   (b)

Figure 8.8 The strokes to surfaces interaction allow users to manually convert strokes into surfaces. (a) diagram of the interaction [51], and (b) interaction on Lift-Off [51].

them more precise but do not follow the conventions and standards of desktop-based CAD. Mixed interfaces also rely on automation to remove the cognitive load of the user. The resulting sketch can be an intermediate step between creating a concept with pen and paper and a 3D model with a CAD tool. Designers of 3D sketching interactions of mixed interfaces need to focus on creating a balance between the need to be precise and how much they constrain users, as they might run the risk of creating an interface that limits the user's creativity [48, 49].

Strokes to surfaces

There are different interactions that achieve a balance between precision and fluidity. One of such interactions is strokes to surfaces [50, 51, 13]. These user interfaces allow users to create 3D strokes with different methods like freehand 3D drawing [13] or tape drawing [50]. In the tape drawing technique, users draw sketches on large scale surfaces using tape. Users unroll the tape with one hand and slide the other hand along with the tape while fastening it on the surface. Then, users extrude these strokes to create surfaces. For the 3D variation, users then extrude these strokes to create surfaces. See Figure 8.8(a) for a diagram of the interaction. An advantage of the stroke to surfaces interaction is that users control both the stroke and the surface creation process. However, users might make errors when sketching the strokes that they will only notice once they extrude them. To solve this problem, Lift-Off [51], illustrated in Figure 8.8(b) automatically creates strokes using a 2D sketch. Users then position these strokes in space before extruding them.

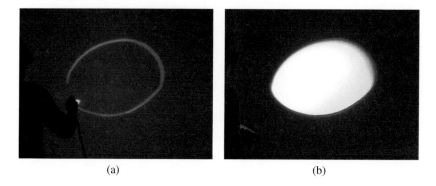

(a)   (b)

Figure 8.9   The 3D strokes to shapes interaction automatically convert the user strokes into 3D shapes. For example, in ImmersiveFiberMesh [53] (a) the user draws a stroke, and (b) the system automatically creates a 3D mesh.

3D strokes to shapes

Other interfaces use the 3D strokes to shape interaction. In these interfaces, users draw strokes in space, and then the system automatically transforms these strokes into 3D models [52, 44, 53, 26] as shown in Figure 8.9. As with the stroke surfaces interaction, different systems use various ways to create strokes. For example, 3D freehand drawing [52, 53] and hand gestures [44]. The difficulty of using the 3D strokes to 3D shape interaction is to develop interpretative algorithms that are not overly sensitive to 3D input inaccuracies. One way to improve user accuracy is to make users draw the strokes in a physical object. For example, a tracked object that allows users to directly create 3D strokes [26].

2D strokes to shapes

A similar interaction technique that also uses physical objects is called 2D strokes to shapes [54, 55, 56, 51, 57, 58]. In this interaction technique, users draw 2D objects that they later can extrude into 3D shapes in a similar way to how some 3D CAD tools work [59]. For example, Mockup Builder [54, 55] shown in Figure 8.10 uses a touch table where users draw strokes. Then users use hand gestures to extrude the new surface. The user's action mimics pulling and pushing the shape out of the table, so it is easy to understand. This interaction technique also allows users to see the result of their extrude in real-time. User interfaces that use the "2D strokes to shapes"

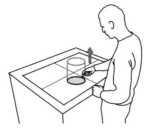

Figure 8.10   The 2D strokes to shapes interaction allow users to manually extrude 2D strokes to create shapes. A system that uses this interaction is (a) Mockup Builder [54].

interaction techniques usually include tools to edit the sketch. For example, some user interfaces allow users to change the position of vertices, edges, and surfaces [57]. Finally, other user interfaces edit sketches by over-sketching the 3D model [58].

Beautification

The last interaction technique in this category is beautification, which translates the user's informal strokes into structured shapes. This interaction technique automatically predicts the user's intended drawn object to transform the stroke into the correct shape. Beautification is widely explored in 2D systems [60, 61, 62, 63], but less in systems for 3D sketching [64, 65, 66]. For example, Multiplanes [64] (Figure 8.11) beautifies the user's stroke to lines and circles in real-time. Doing the beautification in real-time allows users to visualize the detected shape before they finish drawing, which allows users to change their hand movements accordingly. On the technical aspect, the main problem with beautification is developing an algorithm that correctly identifies the user's intention. Otherwise, the user can get frustrated. However, new research on improved object recognition with neural networks [67, 68] is trying to solve this problem. On the interaction aspect, the problem with beautification is that beautified sketches can lose some of their authenticity and stroke expressiveness, for example, lines not meeting at corners or lines which are not straight [69]. Another problem with excessively precise sketches is that it violates the principle of a draft, that "preliminary ideas should look preliminary" [70, 71].

Designers of mixed interfaces for 3D sketching should consider the level of assistance the user interface provides. For example, one decision designers need to make is if the user interface translates the user's strokes

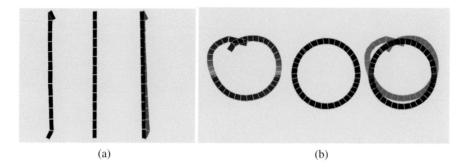

(a)   (b)

Figure 8.11   Beautification examples from Multiplanes [64]. (a) Lines, and (b) circles.

to a clean, well-connected curve network, like FreeDrawer [13] or not, like Multiplanes [64]. Another decision is the amount of control the user has. For example, if the user explicitly controls the extrusion paths like in Lift-Off [51] or chooses from a fixed set of templates like in Window-Shaping [56]. All these decisions affect the type of sketch the user can create.

Finally, regardless of the sketch type, the user interface creates, some design decisions depend on the type of device and interaction techniques used to draw the sketch, as different devices and interaction techniques provide different affordances. In the next section, we will discuss this topic in-depth.

8.2 Interaction Techniques and Devices Affordances

Affordances are an attribute of the tool that shows which features they offer to the user [5, 6]. Therefore, depending on the interaction technique and the device used to create a sketch, their affordances affect the final sketch. For example, it is important to consider the tip width of a pen, as this will affect the level of detail of the sketch. Another feature to consider is the pen shape and size. Designers of user interfaces for 3D sketching need to consider the affordances of the interaction technique and the device they choose. In this section, we focus on the functional and physical affordances described by [5].

8.2.1 Functional Affordances

Functional affordances are a design feature of an interaction technique or device that helps users accomplish their work and are related to the way the tool works [5, 72]. For example, the grid in graph paper provides designers with a functional affordance, as the presence and size of the grid affect the way the user draws. Interaction devices and techniques for 3D sketching have two main functional affordances. The first is the way users control the stroke creation process. The second is the additional tools the system provides to the user to help them be more accurate. Next, we are going to describe these two functional affordances of a user interface.

8.2.1.1 3D Stroke Creation

In the last section, we described 3D freehand drawing as an interaction technique where the user creates a stroke by moving their hand in space. However, this interaction technique presents control and precision issues related to the challenges of 3D sketching, see Section 6.2 for more information. Here, we discuss different interaction techniques that affect the way the stroke follows the drawing tool position. These interaction techniques help users draw more accurately, but they also affect the look of the sketch. For example, the stroke length, that is, short versus long, or the line straightness, that is, curvy versus straight strokes. Finally, any of these stroke creation methods can be used in the sketch types discussed in Section 8.1. However, it is important to note that user interfaces created for different sketch types can use any of the stroke-creation interaction techniques we are going to discuss next.

Follow-hand

The most common stroke creation interaction technique is follow-hand. This interaction technique is easy to understand and implement because it directly translates the action of drawing with a pen to 3D sketching as shown in Figure 8.12. It also allows users absolute control over the resulting geometry, as it will follow all their hand movements. Examples of systems using the follow-hand interaction technique are Smart3DGuides [14], SymbiosisSketch [11], and Multiplanes [64] among others [15, 17, 18, 23, 19]. Current commercial systems, like Tiltbrush [73], also use this interaction technique. It is important to consider the amount of control a user has over the stroke, as the sketch type will dictate the requirements. For example, for conceptual sketches, a stroke with uneven twists along the stroke might be desired, but for a final draft, some filtering needs to be applied to the geometry to prevent them. An example of a system that filters the strokes is SymbiosisSketch [11], filters the user's hand movement based on stroke dynamics (drawing velocity). See Section 7.1 for more information. Finally,

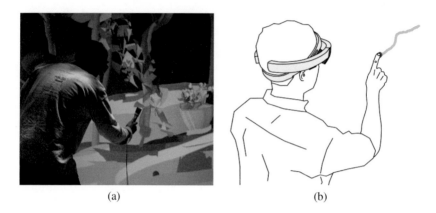

(a)   (b)

Figure 8.12   Follow-hand is the most common stroke creation interaction technique. (a) CavePainting [19], and (b) diagram of the interaction [16].

it is important to consider that too much filtering can limit the characteristic appearance of a sketch as a sketch.

For the follow-hand interaction technique, the type of geometry used to represent the stroke affects the level of control the user has. In user interfaces that use calligraphic stroke, for example ribbons, the arm and the controller orientation affect the stroke, making it difficult to create smooth lines [74]. This problem is less present when the system uses cylinders because then, the controller orientation does not influence the stroke. See Chapter 7 for a technical explanation of surface creation However, regardless of the geometry used, the follow-hand interaction is sensitive to the user's mistakes. Yet, as we mentioned before, this interaction technique also provides users with total control of the stroke in the way of changing the stroke orientation while drawing, which could make it preferred by experts.

3D tape drawing

Another stroke-creation interaction technique based on a 2D drawing technique is 3D tape drawing [18, 50, 74]. 3D tape drawing is a bimanual drawing technique that provides explicit control of the tangent of the drawn curve shown in Figure 8.13. In this interaction technique, the non-dominant hand defines the tangent, and the dominant hand draws the line following the tangent. Some implementations of 3D tape drawing use a drawing plane to project the stroke, as they directly translate the technique from 2D. For

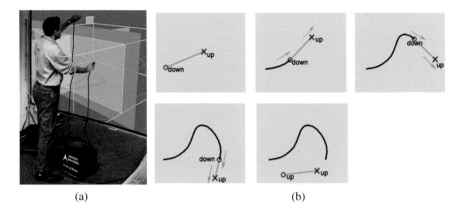

(a)   (b)

Figure 8.13 3D tape drawing uses two hands to give users more control of the drawing stroke. (a) Grossman et al's [50]'s system, and (b) diagram of the interaction [50].

example, virtual planes on Grossman et al.'s system and physical planes on Fleisch et al.'s system. Other implementations, like [74], translate this method to 3D freehand drawing and do not project the stroke to a plane.

Compared to the follow-hand interaction technique, 3D tape drawing helps reduce noise in arm movements. This interaction technique also helps users visualize the curve direction, which translates into better planning of the stroke. However, one problem with 3D tape drawing is that it requires coordination between the two hands. Another problem is that some shapes, such as circles, are difficult to draw. Finally, 3D tape drawing is commonly used when designing objects with curved lines, like cars.

Line-drag

The Line-drag interaction technique [50, 74, 75, 20] can be considered a one hand implementation of the 3D tape drawing interaction technique. When using Line-Drag, the user defines a path that the stroke follows, see Figure 8.14. A metaphor that better describes this interaction technique is dragging a rope with the stroke attached at the end. One characteristic of the line-drag interaction technique is that different implementations apply different rules to the way the stroke follows the line. The simplest implementation is to let users define the rope length. In Grossman et al.'s system, users first define the start point with a click. Then, users draw the path. Finally, with a second click, the stroke follows the drawn path. Another

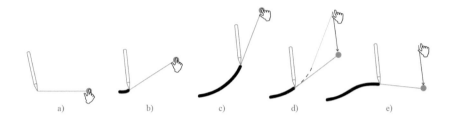

Figure 8.14   Line-drag is a one-hand implementation of 3D tape drawing. Diagram of the interaction [74].

Figure 8.15   The connecting points interaction creates a line between vertices defined by the user. An example of this interaction is Tapline [16].

implementation is to automatically adjust the length of the space between the stroke and the cursor. For example, Drawing on Air [74] uses the stroke length to adjust this length. The system has a predefined minimum and maximum length to avoid errors. On the other hand, Dynamic Dragging [75] adjusts this length based on the expected curvature of the stroke, the drawing speed or a mix of both. A different approach is to use external constraints to affect the length between the stroke and the cursor. For example, Dynasculpt [20] uses a physical simulation that gives the cursor a specific mass, and it considers the path as a damped spring. In this implementation, playing with the cursor and path parameters affect the stroke properties.

The advantages and use cases of the line-drag interaction technique are similar to those of 3D tape drawing. However, line-drag solves the problems of using a bimanual interaction, like the coordination required to draw difficult shapes. The disadvantage of this interaction technique is choosing how the stroke follows the cursor: manual methods can be tedious, but automatic methods can confuse users.

Connecting points

Until now, we discussed stroke creation interaction techniques that emulate drawing with a pen and paper. Other interaction techniques automate the stroke creation process. For example, the connecting points interaction technique automatically creates a line between vertices defined by the user as shown in Figure 8.15. This interaction technique is aimed at drawing straight lines because users only need to define the start and end of the stroke. For drawing curves, a system can use a spline curve, where the stroke interpolates between the points the user gives like Gravity Sketch [76] does.

There are different ways to place the vertices in space. One way is to directly position the vertex in space like Tapline [16] and Brush2Model [45]. Another way is to use an indirect way to position the vertices, like GoGo-Tapline [16], which uses the GoGo-cursor [77] to position the points outside the user's reach. Finally, connecting points can create both strokes [16] and surfaces [44].

Bimanual creation

Another interaction technique that also relies on automation is bimanual creation which allows users to draw using their two hands [51, 22, 45]. Bimanual creation uses two controllers, each attached to one end of the stroke. When using this interaction technique, users define the stroke curvature and length. Then the system translates these values into a stroke. The separation between controllers defines the stroke length, and to change the stroke orientation, users rotate the controllers. Bimanual creation allows users to create smooth curved strokes with many arcs or twists, which can be difficult with other stroke creation interaction techniques as illustrated in Figure 8.16.

Different levels of automation are possible. For example, Lift-Off [51] automatically creates the curves from 2D strokes, and the user only positions and rotates the curve. On the other hand, the system proposed by McGraw et al. [22] uses Hermite spline curves, and the user's wrist orientation controls the curves of the stroke.

Physical-tools

The physical-tools interaction techniques [78, 31, 19] use the physical device shape to communicate the possible actions to the user and try to emulate the real-world capabilities of those devices virtually. For example,

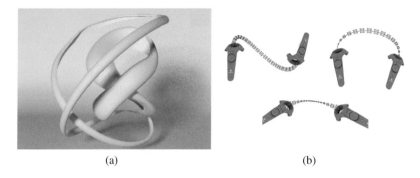

(a)   (b)

Figure 8.16 The bimanual creation interaction uses two controllers to define the curvature and length of a stroke. (a) McGraw et al's [22]'s system, and (b) example of different strokes created by the different combinations of controller positions [22].

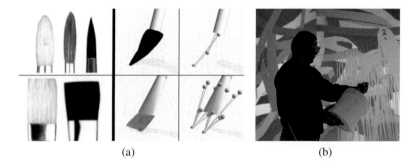

(a)   (b)

Figure 8.17 Examples of systems that use physical-tools. (a) DAB [78] © ACM and (b) CavePainting [19]

emulating the way paint reacts with gravity when thrown from a bucket or the stroke properties of a stroke made by a specific shape of a brush as illustrated in Figure 8.17. It is important to note that the name of these interaction techniques is a metaphor to describe the context the physical device communicates to the user and does not relate to the ergonomic properties of the device. The physical-tools interaction techniques' main advantage is to bring familiar behaviors to 3D sketching. Physical-tools also allows users to explore other design possibilities besides lines and geometrical figures. For example, three brushes of Cave painting [19] enter this category. The Jackson Pollock++ and splat sprinkle a wall of the CAVE with paint, and the Bucket covers a whole surface with paint as if it was thrown with a bucket. Physical-tools also emulate specific tools found in the real-world. Three modelling tools of "Sculpting" [31] enter this category. The Additive Tool leaves a trail of material, like squeezing a toothpaste tube. Second, the Heat Gun melts the existing materials as if it was Styrofoam. And the Sandpaper smooths the shape by wearing away the ridges and filling the valleys. Finally, DAB [78] emulates the feeling of drawing with different brushes. Their 3D haptic brush models also affect the stroke characteristic.

In this section, we presented six stroke-creation interaction techniques with specific advantages and disadvantages. When designing a 3D sketching user interface, designers need to consider which stroke control interaction is the best for their current goal. For example, if users are going to draw objects that consist primarily of straight lines or of curves. However, it is important to remember that users should be able to use the same sketching tool for different situations. As Shneiderman wrote, "design with low thresholds, high ceilings, and wide walls" [79]. In other words, a 3D sketching user interface should be easy to understand, while providing a wide range of functionalities.

8.2.1.2 3D Drawing Tools

In this section, we discuss the functional affordances of the interaction techniques that help users draw more accurately while keeping the fluidity and ease of sketching. These interaction techniques, also called drawing-tools, try to diminish the ergonomic and technological issues of 3D sketching by providing additional drawing tools inside the virtual environment that solve some of the problems mentioned in Section 6.2.2.1. The drawing-tools available in a user interface depend on the type of sketch the interface produces. The type of user also affects the drawing tools available. Finally, to avoid confusing users, it is important to consider the number and type of drawing-tools a user interface provide, as some interaction techniques try to solve the same problem in different ways. Next, we will discuss some popular drawing tools and discuss in which scenarios they are useful.

Rays

Using a ray to show where the user controller intersects with the virtual environment is widely used in virtual reality and augmented reality applications. The goal of the rays interaction technique is to help users know where they are pointing. The same principle applies to 3D sketching interfaces, where a ray shows the exact intersection position of the drawing device with the drawn sketch before they touch it.

(a)   (b)

Figure 8.18 RoMA [38] uses rays to point

The rays interaction technique is useful for editing sketches. For example, in RoMA [38] a ray shows users the part of the sketch they are adjusting as shown in Figure 8.18. In the more modelling-oriented user interfaces, rays also help users identify the selected edge, vertex, or surface. Finally, the rays interaction technique can work together with shape guides and drawing planes to show the position where the stroke is going to be drawn.

Shape-guides

The shape-guides interaction technique consists of having guides inside the virtual environment. These guides help users draw more accurately by allowing them to visualize the errors they make. Most shape guides interaction techniques react to the user's actions, but this is not a requirement. Designers of shape-guide interaction techniques need to consider the shape and functionality of the guide, as these characteristics change depending on the type of sketch the user interface creates.

One shape-guide interaction technique helps users visualize their next stroke [50, 80] by displaying guides a user can trace over. For example, WireDraw [80], shown in Figure 8.19(a), helps users create physical models with a 3D extruder pen. WireDraw needs to know the drawn object so it can divide it into strokes, and order the strokes into steps. After that, while the

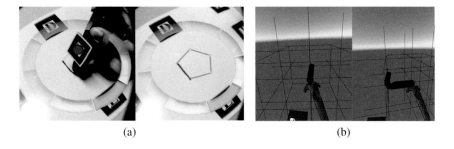

(a)   (b)

Figure 8.19 The shape-guides interaction technique provides users with guides inside the virtual environment. For example, (a) WireDraw [80] and (b) Smart3DGuides [14].

user is drawing, the system shows the position and shape of the next stroke user need to draw. Another shape-guide interaction technique uses a physical object as a guide. For example, Milosevic et al.'s system use a pen to trace over objects that the system automatically translates into 3D models. These shape-guides interaction techniques are useful in situations where the user already knows what they are drawing and wants to transfer the model into another medium. For example, WireDraw requires a digital model to create the physical object. Another use case is when the user has a model and wants to alter it or mix parts of different models. Here using 3D sketching over 3D modelling tools might be beneficial thanks to the properties we have already mentioned, such as fluidity and expressiveness among others. However, it is important to note that tracing virtual objects can make the user's stroke less accurate [24].

Other shape-guide interaction techniques help users orient their drawings in space [14, 28, 27]. For example, in Kim et al.'s system the user creates scaffolds by waving their hands. These scaffolds define the size and general shape of the drawn sketch. Another example is Smart3DGuides [14], shown in Figure 8.19(b), which has local and global reference frames that help users keep the shape-likeness of their sketches consistent. Finally, in SketchingWithHands [28] users use virtual models of their hands in different poses to help them sketch objects that fit in the user's hand. Using shape-guides interaction techniques to help users visualize the final shape assists them in planning their next stroke [14]. However, it is important to avoid filling the virtual environment with guides, as this can be counterproductive, because it can distract or confuse users [82].

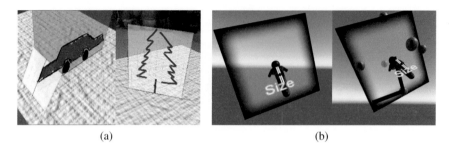

(a)   (b)

Figure 8.20 Drawing-planes give users surfaces in space to draw on. For example, (a) Napkin sketch [83, 84] for AR and (b) Multiplanes [64] for VR.

Drawing-planes

In Section 6.2, we discussed one common error users make when 3D sketching is bending the stroke in depth. The drawing-planes interaction technique helps users prevent this error by projecting the stroke to a plane [64, 15, 28, 27, 29, 38, 83, 84]. In other words, this interaction technique gives users a virtual canvas on which they can draw.

The drawing-planes interaction technique has different implementations depending on the device used to display the virtual environment. In AR headsets, the drawing-planes interaction technique helps extend the real-world to mid-air. For example, Napkin sketch [83], shown in Figure 8.20(a), uses the real-world to create a base plane. Users can define a new canvas by drawing a line on the base-plane to set the orientation of the new plane. Then, the system uses the base-plane normal to create the new plane. Finally, users can further position the new plane by rotating it. In VR headsets, the drawing-planes interaction technique shows a canvas inside the virtual environment. For example, Multiplanes [64], shown in Figure 8.20(b), automatically generates drawing planes using the user's hand orientation and previously created planes. Multiplanes use rules to create parallel, perpendicular, acute angles, and free orientation planes. Another VR example is Hyve-3D [15], where users draw in a tablet that they directly position in the virtual environment. The tablet works like the drawing plane on which the system snaps the stroke.

Finally, the drawing-plane interaction technique is useful for interfaces that use a 2D screen to display the virtual environment [28, 27, 29]. In these devices, users position the plane in the virtual environment using common interactions as in 3D CAD systems, such as widgets. Then, the system

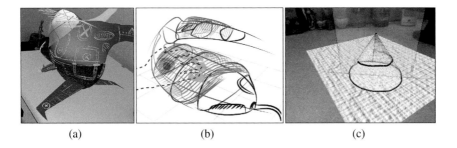

(a)   (b)   (c)

Figure 8.21 Drawing-surfaces give users surfaces of different shapes to draw on. For example, (a) SymbiosisSketch [11], (b) Kim et al's [27]'s system, and (c) Napkin sketch [83, 84].

projects the 2D strokes into these planes, which allows users to create 3D sketches. For example, users translate the drawing-plane using a drag and drop interaction in Mohanty et al.'s system. Users can also define a pivot and a rotation radius to rotate the drawing-plane.

Drawing-surfaces

Like drawing-planes, the drawing-surfaces interaction technique [11, 50, 27, 83, 84] helps users improve their strokes. The difference between both interaction techniques is that the drawing-surfaces interaction technique is not limited to planar surfaces. Instead, users can draw on curved surfaces, including the 3D model of an object. The advantage of drawing-surfaces over drawing-planes is that users have more freedom to create complex shapes.

One example is SymbiosisSketch [11], shown in Figure 8.21(a), where users can draw on previously created strokes. SymbiosisSketch also allows users to create drawing planes and surfaces. Another example of this interaction technique is Kim et al.'s system, shown in Figure 8.21(b), where users draw in scaffolds created by their hand gestures.

In this section, we discussed four different interaction techniques that help users draw more accurately. However, this is not a comprehensive list, as designing drawing-tools that are easy to use and learn is still an open area of research. As with the stroke creation interaction techniques, the best drawing-tool is dependent on the sketch the user is drawing. This decision will also depend on the device used to create the 3D sketch, as the drawing-tool should enhance their affordances.

8.2.2 Physical Affordances

Physical affordances are the characteristics of an interaction device that helps users achieve their tasks. For example, a button needs to have an adequate size and an easy-to-access location [72]. Designers of user interfaces for 3D sketching needs to consider the physical affordances of the input device used to create strokes. It is also important to consider if the user interface is going to provide haptic feedback to the users and the physical affordances this feedback provides. In the following sections, we are going to discuss these two artifacts.

8.2.2.1 Input Device

The pen or pencil a person uses to draw gives different properties to the sketch. For example, the nib used in a calligraphy pen affects the width of the font, and the graphic harness of a pencil tells how hard or soft the stroke is going to be. For 3D sketching user interfaces, the ergonomics of the input device affects user performance [85, 86]. Some of the physical affordances that designers of 3D user interfaces need to consider are the grip type [87], and the weight distribution [88]. Next, we are going to discuss different input devices used for 3D sketching and which affordances they provide.

Controllers

The last wave of VR and AR headsets such as Oculus Rift S [89] and Vive Pro [90] comes with controllers that work for a wide range of applications. See Figure 8.22. For a 3D sketching user interface, these controllers are easy to use, as they are already integrated with the headset [14, 91, 22, 38, 45]. However, most of these controllers do not provide the physical affordances that a user needs to create accurate sketches. For example, most controllers use a power grip that helps users press buttons, but it makes doing precise movements difficult [85]. Designers of user interfaces for 3D sketching can solve the controllers' accuracy issues using functional affordances. However, usually, it is better to use other input devices.

Pens

Most people learn how to use a pen to write and sketch in their infancy, and they can translate this knowledge into 3D sketching. The design of the pen-like device encourages users to hold the pen using their fingers. Zhai et al. [86] found that using the finger muscles to grip the input device has

Figure 8.22 Current controllers for VR. (a) Nolo [92], (b) Windows Mixed Reality Headset [93], (c) Oculus [89], and (d) HTC Vive PRO [90]

better performance than using the wrist or elbows muscles. For example, users can grab the pen using the precision grip, where users hold the pen with their thumb and index fingers, which gives precise control of the movement [87, 94]. Thanks to this affordance many user interfaces for 3D sketching use a pen as an input device [95, 51, 27, 81, 53, 26, 96, 97, 80, 98].

When designing a pen for a 3D sketching system, it is important to consider the way the system tracks the pen. Some devices use markers [96, 97]. For example, ARPen [96], shown in Figure 8.23(a) uses a computer vision algorithm to track pen movement in mid-air. Then using a smartphone users can visualize the drawn sketch. Other user interfaces track the pen using a camera. For example, [81]'s system uses IR-LEDs inside the pen that a camera tracks.

Finally, pens can also provide other physical affordances that enhance the user experience like haptic feedback [95, 8, 98], and force-sensitive sensors [8]. For example, Israel et al.'s pen (Figure 8.23b) allows users to draw by pressing the upper component of the pen until it touches the lower component, and depending on the force by which the user pressed the pen the width of the stroke changes from 1 to 8 mm.

Gestures

Gestures are input events using human hands, and they consist of the movement of fingers and arms in mid-air. One advantage of interactions with gestures is that gesture-based interfaces can reduce the complexity of interaction between humans and computers [100]. And a specific advantage of using gestures for 3D sketching is that this interaction allows for detailed

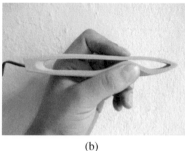

(a)   (b)

Figure 8.23   Pens are common tools used to sketch. For 3D sketching, it is important to consider how the system tracks the pen position in space. Examples of pens for 3D sketching. (a) ARPen [96] and (b) Israel et al's [8] and Zöllner's [99]'s pen.

control using the index finger, and more dynamic control using the whole arm as shown by [101]. See Figure 8.24.

However, when designing a user interface for 3D sketching, it is important to distinguish between manipulative and communicative gestures [102]. Manipulative gestures are those used to interact with objects, for example, take an object, move it somewhere and release it. These gestures do not require attention resources, as using hands is a natural way to interact with the physical world and objects in it, and therefore do not disturb the sketching process. On the other hand, communicative gestures use gestures as symbolic shapes that communicate specific acts. For example, the American sign language (ASL). When using communicative gestures for 3D sketching, users need to remember the meaning behind each sign, which can break the sketching process [41]. Another problem to consider is that current gesture recognition technology is not accurate enough to make the interaction seamless. For example, users can only move their hands after the system detects their gestures. Therefore, even using manipulative gestures can be clumsy, especially as there needs to be a second gesture or input device to tell the system when to start and end a stroke. Finally, gestures can also fatigue the user, known as the gorilla arm effect [103].

For 3D sketching, the gestures interaction technique can use a single hand to interact with the sketch [16, 104, 101, 66]. Usually, these systems use the index finger to draw strokes, for example, Dudley et al.'s system and Virtual Hand [104]. This interaction technique can also use all the fingers to create strokes [44] or the palm [101, 66]. Finally, gestures can use two hands to

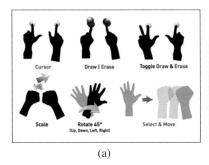
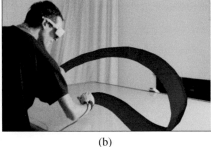

(a)   (b)

Figure 8.24 Gestures allow users to draw strokes using their fingers and arms. For example, (a) users draw with their fingers in AiRSculpt [33], and (b) with their arms in Surface drawing [101].

control the sketch. In these user interfaces, users draw with one hand and edit the sketch using both hands [105, 66, 33]. When evaluating gesture interfaces, previous work found that they present the accuracy problems of 3D sketching [16, 33]. Some users also have problems doing the appropriate gesture to activate an option, which affects the usability of the interaction technique [33].

Finally, there are different ways to track the user's hands and detect their gestures. One way is to use a camera [16, 105, 66, 33], another way is to use a glove [44, 104, 101].

Other-inputs

Other projects have used interesting interaction devices to create strokes that do not emulate using a pen to draw. Here we call them other-inputs. Using other-inputs can bring an element of exploration and creativity to the design process, as the user interface can become part of the process. There are also almost endless possibilities to create physical tools that match particular modelling tasks. However, most of the other-inputs are custom made and require technical knowledge of electronics. Finally, users might find it difficult to use these interaction devices if their shape does not correspond with their functionality. One example of other-inputs is the bucket used to splash paint in CavePainting [19]. Other examples are the tongs and magnets used to edit the sketch in Surface Drawing [101] as shown in Figure 8.25(a). Finally, Leal et al.'s system, shown in Figure 8.25(b) uses

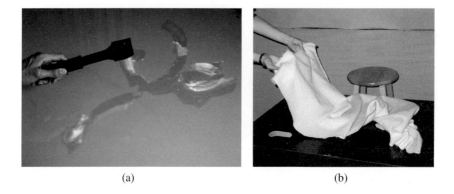

(a)   (b)

Figure 8.25 Other-inputs use physical tools to draw strokes. For example, (a) Surface drawing [101] uses a tong, and (b) Leal et al.'s system uses fabric [32].

fabric to create strokes. This material is flexible to manipulate but strong enough to keep its shape.

The new generation of high-quality VR and AR headsets that have their own controllers made developing 3D sketching user interfaces simpler. Especially as designers do not need to think about the hardware elements. However, in this section, we discussed some of the problems that the current generation of commercial controllers and those made and used in research have. We also presented three other input devices that have their own set of affordances, which help users sketch more accurately. Or in the case of other-inputs, these devices also help the user's creativity. We encourage designers of 3D sketching user interfaces to consider the type of controllers their interface uses.

8.2.2.2 Haptic Feedback
When interacting with objects, humans get a tactile sensation that helps us distinguish different properties of the object, which is known as haptic feedback. However, this sensation is not present when interacting with virtual objects unless we use a device capable of simulating the feeling of touching an object by applying forces, vibrations, or motions [106]. For 3D sketching, haptic feedback helps users improve their accuracy [19, 107]. Next, we discuss different interaction devices and techniques that recreate the feeling of haptic feedback. Each of these interaction devices and techniques has its own set of affordances and problems.

 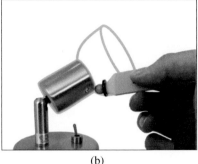

(a)   (b)

Figure 8.26 Drawing over real objects allows users to experience haptic feedback when sketching over them. For example, (a) Jackson et al's [110]'s system and (b) Wacker et al's [35]'s study.

Real objects

Real objects provide passive haptic feedback when the user touches them, and this physical materiality can affect the user's sketch. For example, using a pen to draw provides haptic feedback that affects the way users hold and use the pen to draw. For example, the pen-shaft shape and diameter affect the way users control their movement while sketching [88]. Also, a lightweight pen prevents user fatigue when using the precision grip [85, 108]. Finally, it is this physical materiality that allows users to guide the pen precisely in a precision grip, since slight movements of the fingertips, which are transferred to the pen, result in minimal changes in the position of the pen, which in turn are perceived by the fingertips [109].

It is also possible to use physical 3D models of shapes that users can draw over [110, 81]. For example, Jackson and Keefe [110] use 3D printed shapes of scientific datasets that users can explore by sketching over them as illustrated in Figure 8.26. Using a 3D model as haptic feedback can bring some inaccuracies to the stroke, especially if the tip of the device does not fit in at the edge of a concave or convex shape, as users cannot correctly trace over the outline of the object [35].

Portable surfaces

Portable-surfaces use touch devices to give users a physical surface they can touch. Some user interfaces use a mobile device [56, 11, 15, 83, 18, 111, 112], and other user interfaces use a tablet without a screen [58, 26].

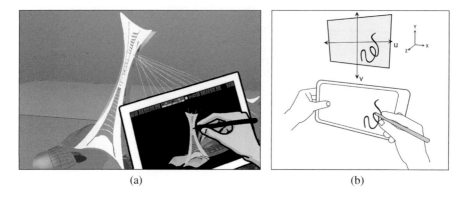

(a)   (b)

Figure 8.27 Examples of systems that use portable surfaces. (a) SymbiosisSketch [11] and (b) Hyve-3D [15]

Portable-surfaces allow users to utilize their 2D drawing skills to sketch in 3D by using the screen as a physical canvas. To position the stroke in place, users point the mobile device to the selected object. For example, SymbiosisSketch [11] uses a tablet to add details to previously drawn objects as shown in Figure 8.27(a). Users need to point the tablet at the object they want to draw on. Another approach is to use the tablet as an interaction device, where users also draw. For example, Hyve-3D [15] uses the tablet as a 3D cursor and as a drawing surface as shown in Figure 8.27(b). Users move the canvas by using touch gestures and draw on the canvas using one finger. In SketchTab3D [112] users 2D sketches using the tablet screen. Then, users position these sketches in space using the tablet. Once a sketch is attached to a tablet, users can continue updating this sketch even if they are not directly looking at it.

One problem with portable-surfaces is that users can not feel the shape of the drawn object, or it's material. Another problem is that in some examples, users need to keep the tablet static while drawing with one hand, which can be tiring. However, one way to avoid this problem is to allow users to move the tablet without moving the sketch, for example, SketchTab3d [112].

Fixed surfaces

Fixed-surfaces are tables or screens that have touch capabilities, as illustrated in Figure 8.28. Users draw on the screen by using their fingers or other objects. This characteristic makes the fixed-surfaces devices have similar

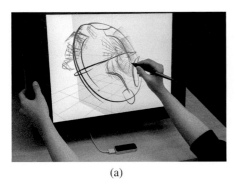
(a)

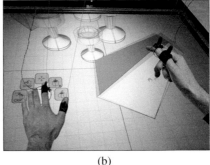
(b)

Figure 8.28 Examples of systems that use fixed surfaces. (a) Kim et al's [27]'s system and (b) Mockup Builder [54]

affordances to the portable-surface devices. One advantage of fixed-surfaces over portable surfaces is that the user does not carry the device, which prevents fatigue.

However, it is important to consider how to change the user's viewpoint when using a fixed-surface device. Some user interfaces use 3D navigation techniques to change the user's viewpoint position [27, 28]. These implementations can be useful to expert designers that have a lot of experience working with 3D CAD systems, as they know how to manipulate the camera fast. However, more novice users can find this approach difficult to use [113]. Other devices use AR headsets to see the shapes outside the screen [54, 55, 57]. These user interfaces are useful for collaborative environments, as many users can see the drawn shapes from multiple viewpoints. Also, most touch tables can support multiple inputs, which helps different designers draw at the same time.

Force feedback

Force-feedback devices use different technologies to manipulate the movement of the device held by the user, see Figure 8.29. For example, vibrotactile actuators [114], pneumatic actuators [115], electrical muscle stimulation (EMS) [116] and mechanical actuators [117]. The way these devices work changes depending on the technology used and a designer of a user interface for 3D sketching needs to choose the correct technology for their system. For example, some devices are fixed in a position, such as The

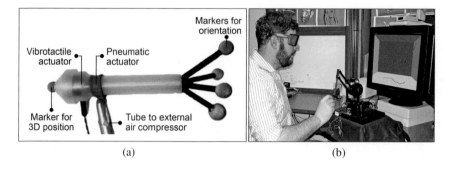

Figure 8.29 Examples of systems that use force-feedback devices. (a) VRSketchPen [98] and (b) Dynamic Dragging [75]

Touch [118] or Phantom [119] among others, while others allow users to move, such as VRSketchPen [98] among others.

Force-feedback devices are intended to feel the size, shape, and properties of different objects. For example, feel the difference when modelling with clay versus wood, or feel the edges of a virtual object. One way to use a force-feedback device is to show the user the position of the drawing canvas [29] or the modeled shape [31]. Another way is to snap the controller to the stroke. For example, Drawing on Air [74] and Dynamic Dragging [75] use 3D tape drawing, and utilize a force-feedback device to constrain the stylus tip to remain on the line segment connecting the two hands. Dynasculpt [20] uses the force-feedback device to alter the sculptural qualities of the stroke. Finally, VRSketchPen [98] allows users to feel the position and the texture of the touched object.

One lesson to take away from this section is that haptic feedback is an important component of any sketching interface. Even the passive haptic feedback of pen-like devices to sketch in VR will affect how the user sketches. When using different types of haptic feedback, it is important to consider how they affect the sketching action. In this section, we also presented other types of passive haptic feedback, like touch surfaces that allow users to utilize their 2D drawing skills to sketch in 3D. Finally, there are also force-feedback devices that create the sensation of touching a virtual object.

8.3 Summary

In this chapter, we present different interaction devices or techniques that help users sketch in 3D and discuss the various advantages and disadvantages of each project. We classified these interaction devices or techniques by the type of sketch, they aim to create (conceptual, technical or a mix). We also discuss the functional and physical affordances each project presents. Our goal was to help designers of future interaction devices or techniques for 3D sketching a general overview of the area so they can design their own interaction devices or techniques.

References

[1] J. F. Hughes, A. van Dam, M. McGuire, D. F. Sklar, J. D. Foley, S. K. Feiner, and K. Akeley. *Computer Graphics: Principles and Practice.* Addison-Wesley, 1996.

[2] C. K. Lim. An Insight Into The Freedom Of Using A Pen: Pen-based System And Pen-and-Paper. In *Proceedings of the Annual Conference of the Association for Computer Aided Design in Architecture (ACADIA '03)*, pages 385–393, 2003.

[3] A. T. Purcell and J. S. Gero. Drawings and the Design Process. *Design Studies*, 19(4):389–430, 1998.

[4] D. G. Ullman, S. Wood, and D. Craig. The Importance of Drawing in the Mechanical Design Process. *Computers and Graphics*, 14(2): 263–274, 1990.

[5] H. R. Hartson. Cognitive, Physical, Sensory, And Functional Affordances In Interaction Design. *Behaviour and Information Technology*, 22(5):315–338, 2003.

[6] D. A. Norman. *The Design of Everyday Things*. Basic Books, Inc., USA, 2002. ISBN 9780465067107.

[7] C. L. Dym. *Engineering Design: A Synthesis of Views*. Cambridge University Press, 1994.

[8] J. H. Israel, E. Wiese, M. Mateescu, C. Zöllner, and R. Stark. Investigating Three-dimensional Sketching For Early Conceptual Design—results From Expert Discussions And User Studies. *Computers & Graphics*, 33(4):462–473, 2009.

[9] J. H. Israel, J. Hurtienne, A. Pohlmeyer, C. Mohs, M. Kindsmüller, and A. Naumann. On Intuitive Use, Physicality And Tangible User Interfaces. *International Journal of Arts and Technology*, 2:348–366, 03 2009.

[10] I. M. Verstijnen, J. M. Hennessey, C. van Leeuwen, R. Hamel, and G. Goldschmidt. Sketching and Creative Discovery. *Design Studies*, 19(4):519–546, 1998.

[11] R. Arora, R. Habib K., T. Grossman, G. Fitzmaurice, and K. Singh. Symbiosissketch: Combining 2D & 3D Sketching for Designing Detailed 3D Objects in Situ. In *Proceedings of the 2018 CHI Conference on Human Factors in Computing Systems*, pages 1–15, 2018.

[12] B. Yee, Y. Ning, and H. Lipson. Augmented Reality In-Situ 3D Sketching of Physical Objects. *Proceedings of the Intelligent UI Workshop on Sketch Recognition*, pages 1–4, 2009.

[13] G. Wesche and H. P. Seidel. FreeDrawer: A Free-Form Sketching System on the Responsive Workbench. In *Proceedings of the ACM Symposium on Virtual Reality Software and Technology (VRST '01)*, page 167, 2001.

[14] M. D. Barrera Machuca, W. Stuerzlinger, and P. Asente. Smart3DGuides: Making Unconstrained Immersive 3D Drawing More Accurate. In *25th ACM Symposium on Virtual Reality Software and Technology*, pages 1–13, 2019.

[15] T. Dorta, G. Kinayoglu, and M. Hoffmann. Hyve-3D and the 3D Cursor: Architectural co-design with freedom in Virtual Reality. *International Journal of Architectural Computing*, 14(2):87–102, 2016.

[16] J. J. Dudley, H. Schuff, and P. O. Kristensson. Bare-Handed 3D Drawing in Augmented Reality. In *Proceedings of the ACM Conference on Designing Interactive Systems (DIS '18)*, pages 241–252, 2018.

[17] S. Eroglu, S. Gebhardt, P. Schmitz, D. Rausch, and Torsten W. K. Fluid Sketching—Immersive Sketching Based on Fluid Flow. In *2018 IEEE Conference on Virtual Reality and 3D User Interfaces (VR)*, pages 475–482, 2018.

[18] T. Fleisch, G. Brunetti, P. Santos, and A. Stork. Stroke-input Methods For Immersive Styling Environments. *Proceedings - Shape Modeling International SMI 2004*, pages 275–283, 2004.

[19] D. F. Keefe, D. A. Feliz, T. Moscovich, D. H. Laidlaw, and J. J. LaViola. CavePainting: A Fully Immersive 3D Artistic Medium and Interactive Experience. In *Proceedings of the 2001 Symposium on Interactive 3D Graphics*, I3D '01, pages 85–93, 2001.

[20] S. Snibbe, S. Anderson, and B. Verplank. Springs and Constraints for 3D Drawing. In *Proceedings of the Third Phantom Users Group*, 1998.

[21] S. Tano, N. Kanayama, T. Hashiyama, J. Ichino, and M. Iwata. 3D Sketch System Based On Life-sized And Operable Concept Enhanced By Three Design Spaces. *Proceedings - 2014 IEEE International Symposium on Multimedia, ISM 2014*, pages 245–250, 2015.

[22] T. McGraw, E. Garcia, and D. Sumner. Interactive Swept Surface Modeling In Virtual Reality With Motion-tracked Controllers. In *Proceedings of the EUROGRAPHICS Symposium on Sketch-Based Interfaces and Modeling (SBIM '17)*, pages 1–9, 2017.

[23] Y. Kim, B. Kim, J. Kim, and Y. J. Kim. CanvoX: High-resolution VR Painting in Large Volumetric Canvas, 2017.

[24] R. Arora, R. H. Kazi, F. Anderson, T. Grossman, K. Singh, and G. Fitzmaurice. Experimental Evaluation of Sketching on Surfaces in VR. In *Proceedings of the 2017 CHI Conference on Human Factors in Computing Systems*, CHI '17, page 5643–5654, 2017.

[25] M. D. Barrera Machuca, W. Stuerzlinger, and Paul Asente. The Effect of Spatial Ability on Immersive 3D Drawing. In *Proceedings of the 2019 on Creativity and Cognition*, C&C '19, page 173–186, 2019.

[26] M. A. Schroering, C. M. Grimm, and R. Pless. A New Input Device for 3D Sketching. *Vision Interface*, pages 311–318, 2003.

[27] Y. Kim, S. G. An, J. H. Lee, and S. H. Bae. Agile 3D Sketching with Air Scaffolding. In *Proceedings of the SIGCHI Conference on Human Factors in Computing Systems (CHI '18)*, pages 1–12, 2018.

[28] Y. Kim and S. H. Bae. SketchingWithHands: 3D Sketching Handheld Products with First-Person Hand Posture. In *Proceedings of the 29th Annual Symposium on User Interface Software and Technology*, pages 797–808, 2016.

[29] R. R. Mohanty, Umema H. B., E. Ragan, and Vinayak. Kinesthetically Augmented Mid-Air Sketching of Multi-Planar 3D Curve-Soups. In *International Design Engineering Technical Conferences & Computers and Information in Engineering Conference (ASME 2018)*, pages 1–11, 2018.

[30] D. Donath and H. Regenbrecht. Using Virtual Reality Aided Design Techniques For Three-dimensional Architectural Sketching. In *Design Computation, Collaboration, Reasoning, Pedagogy. ACADIA Conference Proceedings*, pages 201–212, 1996.

[31] T. A. Galyean and J. F. Hughes. Sculpting: An Interactive Volumetric Modeling Technique. *Proceedings of the 18th Annual Conference on Computer Graphics and Interactive Techniques, SIGGRAPH 1991*, 25: 267–274, 1991.

[32] A. Leal, L. Schaefer, D. Bowman, F. Quek, and C. Stiles. 3D Sketching Using Interactive Fabric For Tangible And Bimanual Input. *Proceedings - Graphics Interface*, pages 49–56, 2011. ISSN 07135424.

[33] S. A. Jang, H. Kim, W. Woo, and G. Wakefield. AiRSculpt: A Wearable Augmented Reality 3D Sculpting System. In *Distributed, Ambient, and Pervasive Interactions, Lecture Notes in Computer Science*, volume 8530, pages 130–141. Springer International Publishing, 2014.

[34] F. L. Krause and J. Lüddemann. *Virtual Clay Modelling*, pages 162–175. Springer US, Boston, MA, 1997.

[35] P. Wacker, A. Wagner, S. Voelker, and J. Borchers. Physical Guides: An Analysis of 3D Sketching Performance on Physical Objects in Augmented Reality. In *Proceedings of the Symposium on Spatial User Interaction*, SUI '18, page 25–35, 2018.

[36] Autodesk. Autocad. https://www.autodesk.com/products/autocad/over view, 2020.

[37] Dassault Systems. Solidworks. https://www.solidworks.com/, 2020.

[38] H. Peng, J. Briggs, C. Y. Wang, K. Guo, J. Kider, S. Mueller, P. Baudisch, and F. Guimbretière. Roma: Interactive fabrication with augmented reality and a Robotic 3D printer. *Conference on Human Factors in Computing Systems - Proceedings*, 2018-April:1–12, 2018.

[39] M. F. Deering. The HoloSketch VR Sketching System. *Communications of the ACM*, 39(5):54–61, 1996.

[40] F. Bruno, M. L. Luchi, M. Muzzupappa, and S. Rizzuti. A Virtual Reality Desktop Configuration for Free-Form Surface Sketching. In *Proceedings on XIV Congreso Internacional de Ingeniería Gràfica*, 2002.

[41] J. Petruschat. Some Remarks on Drawing. *Form+zweck. How to Handle Hands?*, 18:70–77, 2001.

[42] S. Nam and Y. Chai. SPACESKETCH: Shape Modeling With 3D Meshes And Control Curves In Stereoscopic Environments. *Computers and Graphics (Pergamon)*, 36(5):526–533, 2012.

[43] J. H. Kwon, H. W. Choi, J. I. Lee, and Y. H. Chai. Free-hand stroke based NURBS surface for sketching and deforming 3D contents. In *Advances in Multimedia Information Processing - PCM 2005*, volume 3767 LNCS, pages 315–326, 2005.

[44] M. Fuge, M. E. Yumer, G. Orbay, and L. B. Kara. Conceptual Design And Modification Of Freeform Surfaces Using Dual Shape Representations In Augmented Reality Environments. *CAD Computer Aided Design*, 44(10):1020–1032, 2012.

[45] X. Zhu, L. Song, L. You, M. Zhu, X. Wang, and X. Jin. Brush2Model: Convolution Surface-Based Brushes for 3D Modelling in Head-Mounted Display-Based Virtual Environments. *Computer Animation and Virtual Worlds*, 28(3-4):1–10, 2017.

[46] F. Bruno, M. L. Luchi, M. Muzzupappa, M. and S. Rizzuti. The Over-sketching Technique for Free-hand Shape Modelling in Virtual Reality. In *Proceedings of Virtual Concept*, pages 5-7, 2003.

[47] R. Schmidt, A. Khan, G. Kurtenbach, and K. Singh. On Expert Performance in 3D Curve-Drawing Tasks. In *Proceedings of the EUROGRAPHICS Symposium on Sketch-Based Interfaces and Modeling (SBIM '09)*, pages 133–140, 2009.

[48] S. Lee and J. Yan. The Impact Of 3D Cad Interfaces On User Ideation: A Comparative Analysis Using Sketchup And Silhouette Modeler. *Design Studies*, 44:52–73, 2016.

[49] D. Veisz, E. Z. Namouz, S. Joshi, and J. D. Summers. Computer-aided Design Versus Sketching: An Exploratory Case Study. *Artificial Intelligence for Engineering Design, Analysis and Manufacturing: AIEDAM*, 26(3):317–335, 2012.

[50] T. Grossman, R. Balakrishnan, G. Kurtenbach, G. Fitzmaurice, A. Khan, and B. Buxton. Creating Principal 3d Curves With Digital Tape Drawing. In *Proceedings of the SIGCHI Conference on Human Factors in Computing Systems (CHI '02)*, pages 121–128, New York, New York, USA, 2002.

[51] B. Jackson and D. F. Keefe. Lift-Off: Using Reference Imagery and Freehand Sketching to Create 3D Models in VR. *IEEE Transactions on Visualization and Computer Graphics*, 22(4):1442–1451, 2016.

[52] R. De Amicis, F. Bruno, A. Stork, and M. L. Luchi. The Eraser Pen: A New Interaction Paradigm for Curve Sketching in 3D. *DS 30: Proceedings of DESIGN 2002, the 7th International Design Conference,* pages 465–470, 2002.

[53] H. Perkunder, J. H. Israel, and M. Alexa. Shape Modeling with Sketched Feature Lines in Immersive 3D Environments. In *Proceedings of the EUROGRAPHICS Symposium on Sketch-Based Interfaces and Modeling (SBIM '10)*, pages 127–134, 2010.

[54] B. R. De Araùjo, G. Casiez, and J. A. Jorge. Mockup Builder: Direct 3D Modeling On and Above the Surface in a Continuous Interaction Space. In *Proceedings of the Graphics Interface Conference (GI '12)*, pages 173–180, 2012.

[55] B. R. De Araújo, G. Casiez, J. A. Jorge, and M. Hachet. Mockup Builder: 3D modeling On and Above the Surface. *Computers & Graphics*, 37(3):165 – 178, 2013.

[56] K. Huo, Vinayak, and K. Ramani. Window-Shaping. In *Proceedings of the International Conference on Tangible, Embedded, and Embodied Interaction (TEI '17)*, pages 37–45, 2017.

[57] P. Reipschläger and R. Dachselt. Designar: Immersive 3D-modeling Combining Augmented Reality with Interactive Displays. *ISS 2019 - Proceedings of the 2019 ACM International Conference on Interactive Surfaces and Spaces*, pages 29–41, 2019.

[58] H. Seichter. SKETCHAND+ A Collaborative Augmented Reality Sketching Application. *Proceedings of the 8th International Conference on Computer-Aided Architectural Design Research in Asia*, pages 209–222, 2003.

[59] Trimble. sketchup. https://www.sketchup.com/, 2020.

[60] J. Fišer, P. Asente, and D. Sýkora. ShipShape : A Drawing Beautification Assistant. *Joint Symposium of Computational Aesthetics (CAe) Non-Photorealistic Animation and Rendering (NPAR) Sketch-Based Interfaces and Modeling (SBIM) (Expressive'15)*, 2015.

[61] T. Igarashi, S. Matsuoka, S. Kawachiya, and H. Tanaka. Interactive Beautification: A Technique for Rapid Geometric Design. In *Proceedings of the 10th Annual ACM Symposium on User Interface Software and Technology*, UIST '97, page 105–114, 1997.

[62] G. Orbay and L. B. Kara. Beautification of Design Sketches Using Trainable Stroke Clustering and Curve Fitting. *IEEE Transactions on Visualization and Computer Graphics*, 17(5):694–708, 2011.

[63] G. Johnson, M. D. Gross, J. Hong, and E. Y. Do. Computational Support for Sketching in Design: A Review. *Foundations and Trends in Human–Computer Interaction*, 2(1):1–93, 2009.

[64] M. D. Barrera Machuca, P. Asente, W. Stuerzlinger, J. Lu, and B. Kim. Multiplanes: Assisted Freehand VR Sketching. In *Proceedings of the Symposium on Spatial User Interaction*, SUI '18, page 36–47, 2018.

[65] M. Fiorentino, G. Monno, P. A. Renzulli, and A. E. Uva. 3D Sketch Stroke Segmentation and Fitting in Virtual Reality. In *International Conference on the Computer Graphics and Vision*, pages 188–191, 2003.

[66] S. Shankar and R. Rai. Sketching In Three Dimensions: A Beautification Scheme. *Artificial Intelligence for Engineering Design, Analysis and Manufacturing: AIEDAM*, 31(3):376–392, 2017.

[67] Surbhi Gupta, Munish Kumar, and Anupam Garg. Improved object recognition results using SIFT and ORB feature detector. 78(23): 34157–34171.

[68] C. Federer, H. Xu, A. Fyshe, and J. Zylberberg. Improved Object Recognition Using Neural Networks Trained to Mimic the Brain's Statistical Properties. *Neural Networks*, 131:103–114, 2020.

[69] D. Cooper. Imagination's Hand: The Role Of Gesture In Design Drawing. *Design Studies*, 54:120–139, 2018.

[70] E. Y. L. Do and M. D. Gross. Drawing as a Means to Design Reasoning. In *Artificial Intelligence in Design '96 Workshop on Visual Representation, Reasoning and Interaction in Design*, pages 1–11, 1996.

[71] W. Buxton. *Sketching User Experiences: Getting The Design Right And The Right Design*. Morgan Kaufmann, San Francisco, 2007.

[72] H. Rex and S. P. Partha. *The UX Book*. Morgan Kaufmann, Boston, 2012. ISBN 9780465067107.

[73] Google. Tilt Brush. https://www.tiltbrush.com/, 2020.

[74] D. Keefe, R. Zeleznik, and D. Laidlaw. Drawing on Air: Input Techniques for Controlled 3D line Illustration. *IEEE Transactions on Visualization and Computer Graphics*, 13(5):1067–1081, 2007.

[75] D. F. Keefe and R. C. Zeleznik. Tech-note: Dynamic Dragging for Input of 3D Trajectories. In *2008 IEEE Symposium on 3D User Interfaces*, pages 51–54, 2008.

[76] Gravity Sketch. Gravity Sketch. https://www.gravitysketch.com/, 2020.

[77] I. Poupyrev, M. Billinghurst, S. Weghorst, and T. Ichikawa. The Go-Go Interaction Technique: Non-Linear Mapping for Direct Manipulation in VR. In *Proceedings of the 9th Annual ACM Symposium on User Interface Software and Technology*, UIST '96, page 79–80, 1996.

[78] B. Baxter, V. Scheib, M. C. Lin, and D. Manocha. DAB: Interactive Haptic Painting with 3D Virtual Brushes. *Proceedings of the 28th*

Annual Conference on Computer Graphics and Interactive Techniques, SIGGRAPH 2001, pages 461–468, 2001.

[79] B. Shneiderman. Creativity Support Tools Accelerating Discovery and Innovation. *Communications of the ACM*, 50(12):20–32, 2007.

[80] Y. T. Yue, X. Zhang, Y. Yang, G. Ren, Y. K. Choi, and W. Wang. WireDraw. In *Proceedings of the SIGCHI Conference on Human Factors in Computing Systems (CHI '17)*, pages 3693–3704, 2017.

[81] B. Milosevic, F. Bertini, E. Farella, and S. Morigi. A SmartPen for 3D Interaction and Sketch-based Surface Modeling. *International Journal of Advanced Manufacturing Technology*, 84(5-8):1625–1645, 2016.

[82] B. Shneiderman, C. Plaisant, and M. Cohen. *Designing the User Interface*. Pearson, 6th edition, 2016.

[83] M. Xin, E. Sharlin, and M. Costa Sousa. Napkin Sketch: Handheld Mixed Reality 3D Sketching. In *Proceedings of the 2008 ACM Symposium on Virtual Reality Software and Technology*, VRST '08, pages 223–226, 2008.

[84] M. Xin. *3D Sketching and Collaborative Design with Napkin Sketch*. PhD thesis, University of Calgary, Canada, 2011.

[85] D. M. Pham and W. Stuerzlinger. Is the Pen Mightier than the Controller? A Comparison of Input Devices for Selection in Virtual and Augmented Reality. In *25th ACM Symposium on Virtual Reality Software and Technology*, VRST '19, 2019.

[86] S. Zhai, P. Milgram, and W. Buxton. The Influence of Muscle Groups on Performance of Multiple Degree-of-Freedom Input. In *Proceedings of the SIGCHI conference on Human factors in computing systems*, pages 308–315, 1996.

[87] C. M. Schneck. Comparison of Pencil-Grip Patterns in First Graders with Good and Poor Writing Skills. *The American Journal of Occupational Therapy: Official Publication of the American Occupational Therapy Association*, 45(8):701–706, 1991.

[88] R. S. Goonetilleke, E. R. Hoffmann, and A. Luximon. Effects Of Pen Design On Drawing And Writing Performance. *Applied Ergonomics*, 40(2):292–301, 2009.

[89] Oculus by Facebook. Oculus Rift S: VR Headset for VR Ready PCs. https://www.oculus.com/rift-s/, 2019.

[90] HTC. Vive PRO | The Professional Grade VR Headset. https://www.vive.com/eu/product/vive-pro/, 2018.

[91] H. Gardner, D. Lifeng, Q. Wang, and G. Zhou. Line Drawing In Virtual Reality Using A Game Pad. *AUIC'06: Proceedings of the 7th Australasian User Interface Conference*, 15:177–180, 2006.

[92] Nolo. Nolo X1 | 6DoF VR Set. https://www.nolovr.com/nolo_x1, 2020.

[93] by Acer. Windows Mixed Reality Headset | TAKING YOU TO A NEW REALITY. https://www.acer.com/ac/en/US/content/series/wmr, 2020.

[94] T. H. Falk, C. Tam, H. Schwellnus, and T. Chau. Grip Force Variability And Its Effects On Children's Handwriting Legibility, Form, And Strokes. *Journal of Biomechanical Engineering*, 132(11), 2010.

[95] M. Fiorentino, A. E. Uva, and G. Monno. The SenStylus: A Novel Rumble-Feedback Pen Device for CAD Application in Virtual Reality. *13th International Conference in Central Europe on Computer Graphics, Visualization and Computer Vision 2005*, pages 131–138, 2005.

[96] P. Wacker, O. Nowak, S. Voelker, and J. Borchers. ARPen : Mid-Air Object Manipulation Techniques for a Bimanual AR System with Pen & Smartphone. In *Proceedings of the SIGCHI Conference on Human Factors in Computing Systems (CHI '19)*, pages 1–12, 2019.

[97] P. C. Wu, R. Wang, K. Kin, C. Twigg, S. Han, M. H. Yang, and S. Y. Chien. DodecaPen: Accurate 6DoF Tracking of a Passive Stylus. *UIST 2017 - Proceedings of the 30th Annual ACM Symposium on User Interface Software and Technology*, pages 365–374, 2017.

[98] H. Elsayed, M. D. Barrera Machuca, C. Schaarschmidt, K. Marky, F. Müller, J. Riemann, A. Matviienko, M. Schmitz, M. Weigel, and M. M¨uhlhäuser. VRSketchPen: Unconstrained Haptic Assistance for Sketching inVirtual 3D Environments. In *Proceedings of the ACM Symposium on Virtual Reality Software and Technology (VRST '20)*, 2020.

[99] C. Zöllner. *Entwurf und Gestaltung eines Hybriden Werkzeugs zum Skizzieren im Dreidimensionalen Raum*. Diploma thesis, Hochschule

für Technik und Wirtschaft Dresden, University of Applied Sciences (FH), Dresden, 2007.

[100] T. Vuletic, A. Duffy, L. Hay, C. McTeague, G. Campbell, and M. Grealy. Systematic Literature Review of Hand Gestures Used in Human Computer Interaction Interfaces. *International Journal of Human Computer Studies*, 129(March):74–94, 2019.

[101] S. Schkolne, M. Pruett, and P. Schröder. Surface drawing: Creating Organic 3D Shapes with the Hand and Tangible Tools. In *Proceedings of the SIGCHI Conference on Human Factors in Computing Systems (CHI '01)*, pages 261–268, 2001.

[102] F. Quek, D. McNeill, R. Bryll, S. Duncan, X. F. Ma, C. Kirbas, K. E. McCullough, and R. Ansari. Multimodal Human Discourse: Gesture and Speech. *ACM Transactions on Computer Human Interactions*, 9 (3):171–193, September 2002.

[103] J. D. Hincapié-Ramos, X. Guo, P. Moghadasian, and P. Irani. Consumed Endurance: A Metric to Quantify Arm Fatigue of Mid-Air Interactions. In *Proceedings of the SIGCHI Conference on Human Factors in Computing Systems*, CHI '14, page 1063–1072, 2014.

[104] M. Kavakli and D. Jayarathna. Virtual Hand: An Interface For Interactive Sketching In Virtual Reality. *Proceedings - International Conference on Computational Intelligence for Modelling, Control and Automation, CIMCA 2005 and International Conference on Intelligent Agents, Web Technologies and Internet*, 1(May):613–618, 2005.

[105] J. Huang and R. Rai. Conceptual Three-dimensional Modeling Using Intuitive Gesture-based Midair Three-dimensional Sketching Technique. *Journal of Computing and Information Science in Engineering*, 18(4):1–13, 2018.

[106] G. C. Burdea. *Force and Touch Feedback for Virtual Reality*. John Wiley & Sons, Inc., USA, 1996. ISBN 0471021415.

[107] R. R. Mohanty, R. M. Castillo, E. D. Ragan, and V. R. Krishnamurthy. Investigating Force-Feedback in Mid-Air Sketching of Multi-Planar Three-Dimensional Curve-Soups. *Journal of Computing and Information Science in Engineering*, 20(1), 2020.

[108] A. Gatouillat, A. Dumortier, S. Perera, Y. Badr, C. Gehin, and E. Sejdić. Analysis Of The Pen Pressure And Grip Force Signal

During Basic Drawing Tasks: The Timing And Speed Changes Impact Drawing Characteristics. *Computers in Biology and Medicine*, 87 (May):124–131, 2017.

[109] J. H. Israel. *Hybride Interaktionstechniken des Immersiven Skizzierens in frühen Phasen der Produktentwicklung.* PhD thesis, Technische Universität Berlin, 2010.

[110] B. Jackson and D. F. Keefe. Sketching Over Props: Understanding and Interpreting 3D Sketch Input Relative to Rapid Prototype Props. In *IUI 2011 Sketch Recognition Workshop*, 2011.

[111] K. C. Kwan and H. Fu. Mobi3dsketch: 3D Sketching in Mobile AR. In *Proceedings of the 2019 CHI Conference on Human Factors in Computing Systems*, pages 1–11, 2019.

[112] C. Boddien, J. Heitmann, F. Hermuth, D. Lokiec, C. Tan, L. Wölbeling, T. Jung, and J. H. Israel. SketchTab3d: A Hybrid Sketch Library Using Tablets and Immersive 3D Environments. In *Proceedings of the 2017 ACM Symposium on Document Engineering*, DocEng '17, page 101–104, 2017.

[113] G. Fitzmaurice, J. Matejka, I. Mordatch, A. Khan, and G. Kurtenbach. Safe 3D Navigation. In *Proceedings of the 2008 Symposium on Interactive 3D Graphics and Games*, I3D '08, page 7–15, 2008.

[114] J. Rekimoto. Traxion: A Tactile Interaction Device with Virtual Force Sensation. In *ACM SIGGRAPH 2014 Emerging Technologies*, SIGGRAPH '14, pages 25:1–25:1, 2014.

[115] S. Günther, M. Makhija, F. Müller, D. Schön, M. Mühlhäuser, and M. Funk. PneumAct: Pneumatic Kinesthetic Actuation of Body Joints in Virtual Reality Environments. In *Proceedings of the 2019 on Designing Interactive Systems Conference*, DIS '19, pages 227–240, 2019.

[116] P. Lopes, A. Ion, and P. Baudisch. Impacto: Simulating Physical Impact by Combining Tactile Stimulation with Electrical Muscle Stimulation. In *Proceedings of the 28th Annual ACM Symposium on User Interface Software & Technology*, UIST '15, pages 11–19, 2015.

[117] X. Gu, Y. Zhang, W. Sun, Y. Bian, D. Zhou, and P. O. Kristensson. Dexmo: An Inexpensive and Lightweight Mechanical Exoskeleton for Motion Capture and Force Feedback in VR. In *Proceedings of the 2016*

CHI Conference on Human Factors in Computing Systems, CHI '16, pages 1991–1995, 2016.

[118] by 3D Systems. Touch. https://www.3dsystems.com/haptics-devices/touch, 2020.

[119] by 3D Systems. Phantom. https://www.3dsystems.com/haptics-devices/3d-systems-phantom-premium, 2020.

9

3D Sketching Application Scenarios

Philipp Wacker[1], Rahul Arora[2], Mayra Donaji Barrera Machuca[3], Daniel Keefe[4], and Johann Habakuk Israel[5]

[1]RWTH Aachen University, Germany
[2]University of Toronto, Toronto, Canada
[3]Dalhousie University, Canada
[4]University of Minnesota, US
[5]Hochschule für Technik und Wirtschaft Berlin,
University of Applied Sciences, Germany

The technology needed to enable 3D sketching has become more accessible due to the availability of consumer VR devices and companies have released software that enables users to sketch and design in 3D such as Tilt Brush [1], Gravity Sketch [2], and Quill [3]. Research tools have also been made available for other disciplines to work with such as the Lift-Off and Hyve systems. In the previous chapters of this part, we presented information about how interacting with a 3D sketching system can be realized (Chapter 8) and how the sketched information can be represented internally (Chapter 7). This chapter shifts the focus from the sketching system and the interaction techniques to existing projects in various application domains. We present the created artifacts and use-cases to provide an overview of how different domains *use* mid-air 3D sketching.

After presenting some general findings regarding the advantages and effects of 3D sketching on creativity, we present systems from a variety of different disciplines including art, modelling, movie-making, architecture, visualization design & research, medicine, and cultural heritage.

9.1 Conceptual Design and Creativity

Traditional sketching is a central part of early conceptual design and allows the user to quickly produce initial ideas. Many researchers believe that immersive sketching can be beneficial in this stage as designers can create and see their ideas in 3D without requiring additional sketching techniques. Israel et al. [4] investigated how 3D sketching affects the early conceptual design stages by performing expert discussions and user studies. They concluded several benefits of 3D sketching.

Spatiality: seeing a sketch as a 3D object

1-1 proportions: being able to create and see the object at the same size it could be constructed in

Association: to design and create objects where they could be used or relative to existing objects

Formability: the option to adjust and change virtual models

Israel et al. argue that the biggest advantage is the sketching process itself as it enables the user to recognize the spatiality and spatial thinking involved in creating the 3D sketch. On the other hand, Israel et al. also note that traditional sketching has advantages in other areas so that "sketching on paper will not be replaced".

The Lift-Off System by [5] provides a possibility to combine the two methods. Here, users can first use pen-and-paper sketching to create initial design ideas. These sketches are then placed in VR and the user can lift individual lines from the image into mid-air. Between the lines created this way, surfaces can be generated to create a full model. The authors used their system to create several animal models from sketches such as a moose, a fish, or a lion's head and also let an architect try out this system to design a cabin.

To compare creativity between pen-and-paper sketching and immersive sketching, Yang et al. [6] performed a study in which participants had to generate wearable technology ideas around the model of a human given in the scene or on paper as shown in Figure 9.1. They found out that participants in the VR condition had better focus and better flow while working on the task, indicating greater creativity. Participants in the pen-and-paper condition on the other hand were more relaxed. However, their ideas were more similar to each other compared to the ideas in the VR condition. Yang et al. hypothesize that the ability to view the scene from different viewpoints might lead to a higher amount of novel ideas.

Figure 9.1 Example drawings around a mannequin. Students in the VR condition sketched more novel ideas compared to the pen-and-paper condition. Image is taken from [6].

9.2 Art

The ability to define persisting lines in mid-air is a revolutionary opportunity for the art community. Artists have used various techniques to create the perception of depth in their images but immersive sketching enables new possibilities for 3D art. Using a light pen and long-exposure photography, it has been possible for some time to create an image of strokes performed in 3D. Picasso, for example, used this method to visualize 3D sketches [7]. However, the sketches created this way are always somewhat elusive since the artist can not see the lines while sketching and the static viewing perspective has to be specified beforehand. The immersive sketching systems presented in this chapter allow to create lasting sketches that are visible to the artist, are editable and can be viewed from multiple positions.

Examples of art created with Tilt brush can be seen for example on the website of the "Artist in Residence" program run by Google[1]. Compared with models created for games or for personal fabrication, artistic sketches don't need or even benefit from, the absence of watertight 3D meshes. Instead, the strokes can be used to suggest the form rather than completely specify it, thereby engaging the viewer in the interpretation of the artwork.

A classical distinction regarding art is the perception of art as either the "process" of creating artwork or the finished artifact itself. A third view could be the focus on the perception of the viewer while interacting with the artwork. In the area of immersive sketching, this has for example been made part of the art installation by Rubin and Keefe [9] in their "Hiding Spaces". This VR system features 3D painted mid-air objects in front of virtual

---

[1]https://www.tiltbrush.com/air/

Figure 9.2   Long exposure photograph of Picasso using a light pen to draw lines in mid-air. ©Gettyimages.

backgrounds. As the viewer moves through the exhibit, the background morphs based on the viewing movement of the user making them part of the experience.

Our focus in this chapter, regardless of the interpretation of art—either as the final artifact or as the process of creating the artifact—is the interaction with the immersive sketching system. Therefore, we describe how artists have been *using* immersive sketching systems to create their artwork and do not discuss related areas such as copyright or sampling of digital art. Using an immersive VR sketching system is interesting from an artist's perspective as the artist is leaving the real environment to a large degree, grounding her- or himself in the virtual environment. Chittenden [10] looked at this

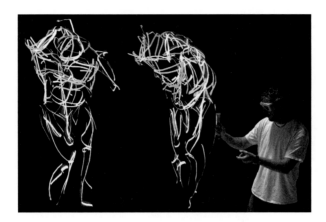

Figure 9.3   An artist suggesting the shape of an artwork by sketching lines in mid-air. Image from [8].

detachment and sees similarities to the "adventure chronotope". Yet, even in this new environment, 3D sketches are unmistakably human creations. In 3D sketching, the human hands can leave a trace that, like a digitally sampled voice in electronic music, provides clear evidence, perhaps even an identifiable artistic style or mark, of the physical art-making process. This has the potential to engage both artists and viewers in new ways with digital media. Indeed, even concepts, such as sampling, derivative works, and copyright take on new meaning when considering a spatial, digital medium that translates the physical movements of the artist so directly. In this regard, 3D sketching is decidedly different from traditional digital 3D modeling tools.

Art can be created in many inventive ways and artists often use existing methods in novel ways to realize their artworks. For example, Grey [11] used the surface drawing system [12] to create a trace around a human person. The resulting model looked nothing like the person but viewing the tracked lines from different perspectives produced interesting shapes. The author used the model in different compositions such as "Rockman" shown in Figure 9.4 and "Visitation". A similar approach is seen in the Artwork by Nam and Keefe [13]. Here, the viewing angle of the person looking at the screen determined how the digital drawing would be shown. Additionally, the artist creating the drawing was recorded from multiple angles. This allowed showing the creation process from these different angles synchronized with the viewing position of the digital drawing while the viewer looks at the artwork.

Figure 9.4   The "Rockman" model created by Grey [11] using the surface drawing system.

There are many ways to create a 2D painting but many early projects for immersive modelling focused on creating cylindrical lines or ribbons in mid-air. To gain more expressiveness and options, Mäkelä [14] added particle clouds and meshes to the mix and also replaced the wand controller with a finger tracking device to easily adjust the size and position of strokes simultaneously in real time. Using this setup, he re-created an image of Rembrandt in VR. More recently, Lisa Padilla[2] used Tilt Brush to create portraits of celebrities such as Stephen Curry, Andy Warhol, or Mark Zuckerberg (Figure 9.5).

To increase the possibilities for mid-air drawing, Eroglu et al. [15] designed techniques for the creation of fluid artworks similar to smoke photography, paper marbling, or ink dripped into water. The "particles" can be manipulated by hand gestures with the controller as shown in Figure 9.6

---

[2]https://www.flickr.com/photos/lisap/44473512285/

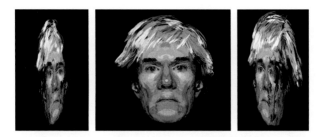

Figure 9.5 Different perspectives at a VR Portrait of Andy Warhol by Lisa Padilla. Created in TiltBrush. Printed with permission.

Figure 9.6 Movements of the hand are tracked to control particles. Images from [15].

or by blowing as shown in Figure 9.7, where the blowing is detected by a microphone that the user is wearing.

Focusing on the experience creating mid-air drawings, Seo et al. [16] presented "Aura Garden", a VR setup in which users can paint their own light sculptures using a custom wand as shown in Figure 9.8. These sculptures remained for the next user to see. While observing visitors to the exhibit, the authors noted that different professions led to different usages of the system.

Figure 9.7 A microphone detects if the user blows air at the particles. Image from [15].

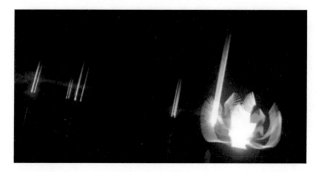

Figure 9.8   Light sculpture created in Aura Garden. Light beacons indicate the location of sculptures of other users. Image from [16].

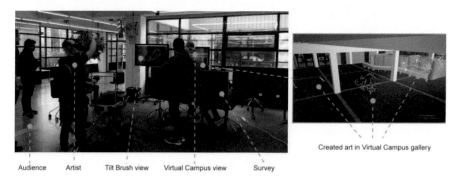

Created art in Virtual Campus gallery

Audience   Artist   Tilt Brush view   Virtual Campus view   Survey

Figure 9.9   Setup of the "Hybrid Campus" project. Both the creation of virtual objects in place as well as the visualization for an audience are co-located on the campus. Image from [17].

For example, photographers created detailed shapes that they would then look at from different angles while dancers enjoyed to visualize their own movement.

In most cases, the creation and consumption of the artworks are separated in both time and space. For their Hybrid Campus Art project, Pakanen et al. [17] combined both of them. Participants could use TiltBrush to create virtual models directly on the University Campus. These models were then placed in a virtual mirror world of the campus and could be viewed on a screen that was located in the same area. The intention of the authors was to also "add value to mirror worlds".

Figure 9.10 Combination of input brushstrokes and the resulting solid model. Image from [19].

9.3 Modeling for Fabrication

While many artists like to only suggest the form of a model through individual strokes as seen in the previous section about art, other domains require solid mesh models. One example is the design of objects that are intended to be printed out using a 3D printer. There are techniques that create 3D models for this purpose for example by using commercially available tools such as Gravity Sketch [2] to create such models directly. In 2007, the Front Design Studio released a video on YouTube showing how they designed furniture by recording mid-air brushstrokes and printing them out using a 3D printer [18]. However, in their system, the designers were not able to see their strokes during the design process. Other projects investigate how the models created by brushstrokes in VR applications such as Tilt Brush [1] can be used to achieve models that can be sent to a 3D printer.

The SurfaceBrush system by Rosales et al. [19] describes an algorithm that interprets the individual brushstrokes in relation to the surrounding brushstrokes to generate surfaces as shown in Figure 9.10. This way, a solid model based on the brushstrokes is generated.

Using a different approach, Giunchi et al. [20] use the user's brushstrokes as search input through a model database. For example, the user creates a rough sketch of a chair and the system searches through its database to find the closest matching existing model. The user can then also adjust the sketch and run the search again.

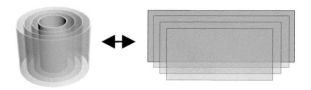

Figure 9.11 The 360 degree surrounding around the VR wearer (left) is separated into different depth layers and flattened into individual planes that can be used to sketch on a tablet. Image from [21].

9.4 Dynamic Storytelling

The previous projects aim to create a model that in its finished state is static. However, 3D sketching can also be used in more dynamic scenarios such as movies and animations. Movies in VR are hard to plan and prototype as traditional planning techniques for camera movement and control have issues when applied to scenarios in which the user controls the viewing direction and can turn 360 degrees at any point. Henrikson et al. [21] address this problem by using 360 degree sketching planes that can be viewed within a VR headset as shown in Figure 9.11. This setup allows for simultaneous work. For example, while the director is wearing the VR headset and watching the scene, an artist can directly implement his comments by adjusting the sketch on a tablet. The director himself can also add notes through a tablet similar to drawing tablets. Different depths in the VR environment are achieved by different 360 degree layers that the user can easily switch between.

Mid-Air interaction can also be used to assist in the creation of 3D animations such as the path of an airplane or wind blowing in a specific pattern. Arora et al. [22] implemented and evaluated the "MagicalHands" system which tracks hand gestures to describe animations. They gave the system to four artists and let them create scenes with animations. The resulting scenes included the setting of scary storytelling, a paper airplane race, and an explanation of how rain is forming. They found out during the use of the system that tracking the coarse hand posture might be sufficient precision for the general interaction so it is not necessary for more precise tracking.

9.5 Architecture

Mid-air sketching also has the potential to be used in different disciplines for which it is not directly apparent. One example is architecture. Traditional

Figure 9.12 Overview of the architectural drawing system. The position of the objects on the mulitouch table is synced with the visualization on the touchscreen. Drawings on the touchscreen are applied to the 3D model depending on the chosen drawing mode. Image from [23].

sketching, physical mockups, and even desktop-based modelling tools, like Google SketchUp, all play a critical role in modern architectural design. 3D sketching complements these by providing options to quickly create a building layout before stepping inside a prototype and annotating the design. For example, the Lift-Off system was used by an architect to design a cabin over multiple sessions with the system [5]. For a dynamic planning scenario, Schubert et al. [23] combined a sketching tool with a physical 3D working model of a cityscape on a multitouch table shown in Figure 9.12. The multitouch table was used to arrange the tangibles representing the buildings and a touchscreen beside the table showed a rendering of the scene. Sketches on this touchscreen could be integrated into the virtual model in different ways. The first, 2D, is similar to tracing on paper. The sketch stays relative to the camera and is not connected to the scene. Adding half a dimension, the 2.5D mode has two sub-modes. In the "Object Mode", lines drawn are projected onto the 3D objects in the scene, in the "Extended Mode", an infinite supporting plane from a selected surface allows to sketch continuations from an already existing plane, for example, to connect two objects or to extend an object. In both 2.5D modes, the drawings are connected to the objects in the scene and keep their relationship when the camera or object is moved. The final mode is the 3D sketching mode. Here, the user can sketch 2D models on the screen and they are converted into 3D mesh models that are placed in the scene as shown in Figure 9.13.

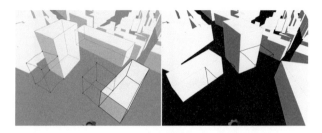

Figure 9.13   Example of the 3D drawing mode. 2D sketches on the screen are converted into 3D models that are placed in the scene. Image from [23].

9.6 Visualization Design and Research

Visualizing information in 3D can be helpful in many disciplines. While artists have the skills to visualize data, interpreting what a set of data means and what data is relevant also requires expert knowledge in the current research domain. Often, artists are employed at the end to create visualizations for the findings of the researchers. However, including visual artists early rather than later during the evaluation can help researchers to *understand* the data and *find* results. Keefe et al. [24] introduced the process of Scientific Sketching to enable the collaboration between researchers and artists to visualize research data. They based this process on the experience they had gathered before when working with visual artists to investigate archaeological dig sites and the fluid dynamic of flying bats by visualizing the data in VR [25]. They found that artists can convey the data well but require critique and assistance from researchers along the way. For the scientific sketching process, they formulate four steps: Paper sketching, VR sketching, VR prototyping, and implementation. Similar to the process suggested in the "Lift-Off" system, the paper sketching step allows for quick exploration of ideas using traditional methods. In the next step, the VR sketching, the ideas can be transferred into the third dimension to both test different spatial arrangements and give people the option to critique these arrangements. The VR prototyping step adds more refinement and interactivity as well as animated elements to the visualization. The resulting visual requirements can now be implemented in the final step. The final visualization is then dependent on the underlying research data and not on manual control. While it is important that the artist can prototype many visualizations in the early stages, it is important to include the real data as well. Otherwise, this might lead to a situation in which the artist has created a good visualization for a single sample of data but this visualization is not able to visualize different

(a)   (b)   (c)

Figure 9.14   The visualization of dinosaur foot movement at different stages of the scientific sketching process. Starting with a paper sketch (a) the artists generate prototype visualization s in VR (b) and suitable visualization s are implemented to run dependent on the research data (c). Images from [26].

sets or the change over time. This process was for example used by Novotny et al. [26] to analyze the foot movement of dinosaurs based on fossil scans. They let two VR visualization design courses work on the visualization and according to the researchers, the process and the result provided a significant benefit to the standard desktop based analysis. To analyze the data, the researchers and artists used different visualizations such as particle visualization to track the movement of different particles during the foot movement as shown in Figure 9.14.

9.7 Medicine

In the medical field, immersive sketching can help in many different ways as well. Similar to modelling tasks described in Section 9.3, seeing the model of new equipment such as an operating table in its real environment can help to uncover issues that have not been detected before. Johnson et al. [27] used the "Lift-Off" system to design new equipment and then have medical professionals look at the models in the real size and setting. This uncovered issues that have not been thought of previously. An example is hinges that are in the way of the doctor when walking around the table. But 3D sketching can be used in more cases such as communication and education. In the same project, Johnson et al. also enabled doctors to create models of injuries such as fractures to explain the diagnosis and treatment to patients as shown in Figure 9.15. The doctors said that while the modelling process itself would not be desirable for them, the option to annotate the model would be used to explain the procedure. An interesting observation of the communication of doctors with the patients was that the doctors spend more time explaining the

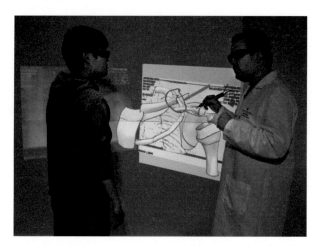

Figure 9.15 A doctor describes the injury and treatment by showing and annotating a 3D model. Image from [27].

surrounding elements of an injury, for example, explaining the effect it had on muscles surrounding a fracture. This included areas that were not even visible in the traditional 2D data normally used in communication with the patient. The researchers conclude that providing 3D models of injuries in their context within the body can help doctors communicate with the patients.

The relationships and processes within the body are hard to learn and the effects of changes are not easy to predict. Saalfeld et al. [28] investigated how 3D sketching can help in understanding such relationships by designing a semi-immersive system with 6-DoF sketching capabilities that is aimed at medical educators and students to understand the vascular structure. Users can sketch blood vessels and add disturbances such as saccular aneurysms. The system would then animate the blood flow through these vessels enabling the teacher or student to In an evaluation, the researchers focused on the creation of blood vessel sketches and they did not study if and how the new visualization improves learning retention.

9.8 Cultural Heritage

Virtual environments have great potential in the area of cultural heritage since it is possible to visualize environments and structures that do not exist anymore. Virtual environments can also be used to provide information and annotate specific areas on artworks without having to mark the statue

Figure 9.16 User can navigate the virtual space using a tablet and annotations made on the tablet are added to the 3D model. Image from [29].

Figure 9.17 The viewing position of a user is shown with a cursor in the 3D model (left). Right: the view of the user from the cursor on the left. Images from [29].

or painting itself. Making the possibility to annotate models available for everyone also creates new ways to involve people in generating cultural heritage. Schnabel et al. [29] used the Hyve3D system and its sketching capabilities (explained more closely in Section 8.2.2.2), to navigate through a scan of the old town of Kashgar. Multiple users could sketch on their tablets to annotate their current view in the 3D world, as shown in Figure 9.16 and Figure 9.17, for example, to provide explanations of interesting structures or 'errors' in the scanning data. The authors explain that this process could enable the creation of new stories either about the old town or its scan in particular.

Figure 9.18   VR sketch of the house of the play "Draw Me Close". *Source:* https://www.nf b.ca/interactive/draw_me_close_en/

Virtual reality has also become an interesting part of theatre installations [30] and mid-air sketching can be used in this context to prototype and design sets. The virtual rooms of the play "Draw Me Close"[3] for example, have been created using mid-air sketching techniques as shown in Figure 9.18.

9.9 Summary

The projects show that 3D sketching can be used in a variety of domains and for many different use cases. One central intention is to "create" a 3D object or scene that can be used later, for example for 3D printing or as an artwork. Other potential uses of 3D sketching are communication and education. Being able to highlight a particular part in a scene or visualize a 3D path can help to bring points across, for example, in medical contexts among others, and being able to manipulate 3D models and see the effects live can assist in understanding the consequences of these changes. We believe that the potential of using 3D sketching as a tool has not yet been reached and that many processes can be improved by employing 3D sketching techniques.

9.10 Conclusion

We conclude for this whole part of the book that 3D sketching is a uniquely powerful and widely applicable tool for designers. Although it is often marketed today as a new technology, taking into account the long history of

---

[3]https://www.nfb.ca/interactive/draw_me_close_en/

3D Sketching, as presented in Section 6.1, reveals a, perhaps more accurate, assessment that what we are witnessing today is instead the maturing of a technology that has been under development for more than three decades.

The longstanding interest in 3D sketching can be attributed, in part, to the many opportunities the hardware and software provide. In other words, the affordances provided by 3D tracking and spatial displays lead naturally to a medium that simultaneously combines an ability to make body-centric 3D spatial judgments with the fluidity, immediacy, and creativity of traditional 2D sketching.

With this fundamental potential identified, it is worth noting that the medium presents its own challenges which also need to be overcome. At the top of this list, presented in Section 6.2, is controlling 3D mark making, a challenging topic because the applications of 3D Sketching are so broad. The most straightforward application of the technologies leads to a style of loose, fluid sketching that is a natural fit for 3D gesture sketches and early conceptual design work. Yet, this is not as appropriate for more deliberate styles of drawing or model making where artists and designers wish to have precise control over proportion, line quality, and form. Fortunately, there are many possible solutions to achieving this level of control as described in the deep dive into interaction devices and techniques together with input processing algorithms in Chapter 8. Compiling and discussing this picture of the state of the art leads us to conclude on behalf of the 3D Sketching research community that we have now reached the point where the goal should no longer be to create the "best" general 3D Sketching system or interaction technique. Instead, the tools and techniques have reached a point of maturity where we are fortunate to be able to instead focus on, "the best technique for the job at hand".

This brings us to the many possible applications of 3D sketching. We hope that our review of compelling work to date inspires designers to think creatively about how 3D Sketching in Space might be used in their practice. The cornerstone application for many designers will likely be conceptual design and modelling for fabrication or building. However, also consider the role 3D sketching can play in art, dynamic storytelling, data visualization , medicine, and cultural heritage, where we see some of the most creative applications to date. Whatever the application, it is clear that 3D sketching in space is now in a renaissance period, and we look forward to seeing how designers will employ these exciting techniques now that it is possible for them to move beyond the research lab and into regular practice.

References

[1] Google. Tilt Brush. https://www.tiltbrush.com/, 2020.

[2] Gravity Sketch. Gravity Sketch. https://www.gravitysketch.com/, 2020.

[3] Facebook. Quill. https://quill.fb.com/, 2020.

[4] J. H. Israel, E. Wiese, M. Mateescu, C. Zöllner, and R. Stark. Investigating Three-dimensional Sketching For Early Conceptual Design—results From Expert Discussions And User Studies. *Computers & Graphics*, 33(4):462–473, 2009.

[5] B. Jackson and D. F. Keefe. Lift-Off: Using Reference Imagery and Freehand Sketching to Create 3D Models in VR. *IEEE Transactions on Visualization and Computer Graphics*, 22(4):1442–1451, 2016.

[6] X. Yang, L. Lin, P. Y. Cheng, X. Yang, Y. Ren, and Y. M. Huang. Examining Creativity Through A Virtual Reality Support System. *Educational Technology Research and Development*, 66(5):1231–1254, 2018.

[7] Pablo Picasso. Pablo Picasso Draws With Light: The Story Behind an Iconic Photo, 2012.

[8] D. F. Keefe. From Gesture To Form: The Evolution Of Expressive Freehand Spatial Interfaces. *Leonardo*, 44(5):460–461, 2011. Publisher: MIT Press.

[9] C. B. Rubin and D. F. Keefe. Hiding Spaces: A CAVE Of Elusive Immateriality. In *ACM SIGGRAPH 2002 conference abstracts and applications on - SIGGRAPH '02*, page 192, 2002.

[10] T. Chittenden. Tilt Brush Painting: Chronotopic Adventures In A Physical-virtual Threshold. *Journal of Contemporary Painting*, 4(2): 381–403, October 2018.

[11] J. Grey. Human-Computer Interaction In Life Drawing, A Fine Artist's Perspective. In *Proceedings of the Sixth International Conference on Information Visualisation*, pages 761–770, 2002.

[12] S. Schkolne, M. Pruett, and P. Schröder. Surface drawing: Creating Organic 3D Shapes with the Hand and Tangible Tools. In *Proceedings of the SIGCHI Conference on Human Factors in Computing Systems (CHI '01)*, pages 261–268, 2001.

[13] J. Nam and D. F. Keefe. Spatial Correlation: An Interactive Display of Virtual Gesture Sculpture. *Leonardo*, 50(1):94–95, February 2014.

[14] W. Mäkelä. Working 3D Meshes and Particles with Finger Tips, towards an Immersive Artists' Interface. In *Proceedings of the IEEE Virtual Reality Workshop*, pages 77–80, 2005.

[15] S. Eroglu, S. Gebhardt, P. Schmitz, D. Rausch, and Torsten W. K. Fluid Sketching—Immersive Sketching Based on Fluid Flow. In *2018 IEEE Conference on Virtual Reality and 3D User Interfaces (VR)*, pages 475–482, 2018.

[16] J. H. Seo, M. Bruner, and N. Ayres. Aura Garden: Collective and Collaborative Aesthetics of Light Sculpting in Virtual Reality. In *Extended Abstracts of the 2018 CHI Conference on Human Factors in Computing Systems - CHI '18*, pages 1–6, 2018.

[17] M. Pakanen, P. Alavesa, H. Kukka, P. Nuottajärvi, Z. Hellberg, L. M. Orjala, N. Kupari, and T. Ojala. Hybrid Campus Art: Bridging Two Realities Through 3D Art. In *Proceedings of the 16th International Conference on Mobile and Ubiquitous Multimedia - MUM '17*, pages 393–399, 2017.

[18] FRONT. Sketch furniture by front. https://www.youtube.com/watch?v =8zP1em1dg5k, 2007.

[19] E. Rosales, J. Rodriguez, and A. Sheffer. SurfaceBrush: From Virtual Reality Drawings to Manifold Surfaces. *ACM Transactions on Graphics*, 38(4):1–15, July 2019.

[20] D. Giunchi, S. James, and A. Steed. 3D Sketching for Interactive Model Retrieval in Virtual Reality. In *Proceedings of the Joint Symposium on Computational Aesthetics and Sketch-Based Interfaces and Modeling and Non-Photorealistic Animation and Rendering*, Expressive '18, 2018.

[21] R. Henrikson, B. Araujo, F. Chevalier, K. Singh, and R. Balakrishnan. Multi-Device Storyboards for Cinematic Narratives in VR. In *Proceedings of the 29th Annual Symposium on User Interface Software and Technology - UIST '16*, pages 787–796, 2016.

[22] R. Arora, R. H. Kazi, D. M. Kaufman, W. Li, and K. Singh. MagicalHands: Mid-Air Hand Gestures for Animating in VR. In *Proceedings of the 32nd Annual ACM Symposium on User Interface Software and Technology*, pages 463–477, 2019.

[23] G. Schubert, E. Artinger, V. Yanev, F. Petzold, and G. Klinker. 3D Virtuality Sketching: Interactive 3D-sketching Based On Real Models In A Virtual Scene. In Mark Johnson Jason Kelly Cabrinha and Kyle Steinfeld, editors, *Proceedings of the 32nd Annual Conference of the Association for Computer Aided Design in Architecture (ACADIA)*, volume 32, pages 409–418. The Printing House Inc, WI, 2012.

[24] D. F. Keefe, D. Acevedo, J. Miles, F. Drury, S. M. Swartz, and D. H. Laidlaw. Scientific Sketching for Collaborative VR Visualization Design. *IEEE Transactions on Visualization and Computer Graphics*, 14(4):835–847, July 2008.

[25] D. F. Keefe, D. B. Karelitz, E. L. Vote, and D. H. Laidlaw. Artistic Collaboration In Designing VR Visualizations. *IEEE Computer Graphics and Applications*, 25(2):18–23, March 2005.

[26] J. Novotny, J. Tveite, M. L. Turner, S. Gatesy, F. Drury, P. Falkingham, and D. H. Laidlaw. Developing Virtual Reality Visualizations for Unsteady Flow Analysis of Dinosaur Track Formation using Scientific Sketching. *IEEE Transactions on Visualization and Computer Graphics*, 25(5):2145–2154, May 2019.

[27] S. Johnson, B. Jackson, B. Tourek, M. Molina, A. G. Erdman, and D. F. Keefe. Immersive Analytics for Medicine: Hybrid 2D/3D Sketch-Based Interfaces for Annotating Medical Data and Designing Medical Devices. In *Proceedings of the 2016 ACM Companion on Interactive Surfaces and Spaces - ISS Companion '16*, pages 107–113, 2016.

[28] P. Saalfeld, A. Stojnic, B. Preim, and S. Oeltze-Jafra. Semi-Immersive 3D Sketching of Vascular Structures for Medical Education. In *Eurographics Workshop on Visual Computing for Biology and Medicine*, pages 123–132, 2016.

[29] M. A. Schnabel, S. Aydin, T. Moleta, D. Pierini, and T. Dorta. Unmediated Cultural Heritage via Hyve-3D: Collecting Individual and Collective Narratives with 3D Sketching. In *Proceedings of the*

21st International Conference of the Association for Computer-Aided Architectural Design Research in Asia CAADRIA 2016, volume 21, pages 683–692, 2016.

[30] S. Atkinson and H. W. Kennedy. Extended Reality Ecosystems: Innovations in Creativity and Collaboration in the Theatrical Arts. *Refractory: A Journal of Entertainment Media*, 30(2018), 2018. ISSN 1447-4905.

10

Conclusions

Alexandra Bonnici and Kenneth P. Camilleri

University of Malta, Malta

In this book, we presented a holistic picture of sketch-based interactions from two-dimensional sketching in physical and digital ink to sketch interactions in the three-dimensions (3D) domain, presenting the technologies and algorithms needed to tackle the interpretation and use of sketches as well as discussing open issues and limitations of existing techniques. We started the book by noting that there are different types of sketching styles that are used throughout the design process, starting with simple, fast concept sketches which serve for design exploration and ideation, to more detailed technical drawings. We note that drawings are used as a means of communication between clients, designers, and peers and that creating three-dimensional prototypes rapidly and efficiently are advantageous and necessary to idea exploration, giving different application examples and scenarios in Chapter 9.

In Chapter 2 we focus on the processing and understanding of paper-based sketches. We observe that some degree of sketch simplification is required to overcome the roughness of the drawings, particularly when taking into account early-stage concept sketches. Here we note the differences between traditional feature-based approaches and, the more recent, deep-learning approaches. While the former focus on the hand-crafting of image-based feature filters such as Gabor filters [2] or Pearson correlations [7], deep-learning approaches are example-driven and, therefore, require sketch and ground-truth data pairs to provide the training examples necessary for the convolutional neural network to learn the simplification kernels.

In Chapter 4, we note that deep-learning approaches have also had an impact on the 3D reconstruction problem, whose solution has seen shifts from the planar-equation style interpretation to a reconstruction-by-example

263

approach, using the convolutional kernels to learn the mapping between two-dimensions (2D) and 3D. The success of these approaches depends on the availability of sketch and 3D object pairs. The NPR techniques discussed in Chapter 5 are useful in this regard, using the NPR techniques to render 3D objects as sketches, thereby generating the required training data. The methods presented in Chapter 5 also present an alternative route to obtaining 3D models from sketches, mainly through retrieving models that correspond to the sketched input.

The second part of the book focuses on sketching directly in the 3D domain. Similar to 2D online sketch-based interactions, sketching in the 3D domain requires the use of gestures to allow the user to interact with the sketch, and, in the 3D domain too, these gestures should be as natural as possible, allowing for seamless integration with the natural gestures the users have around 3D objects in the real world. This becomes necessary so as not to overload users with different gestures and their meanings when working in immersive environments. Unlike 2D online sketching, where the most commonly used drawing tool is the stylus pen, sketching in 3D can use different hand-held controllers which allow the VR/AR system to track the user's gestures in 3D space. Moreover, the user is often required to wear some form of the headset which permits the visualisation of the rendered sketch. After an exposition of the 3D drawing materials, interactions and drawing processing techniques, Chapter 9 discusses the application areas for 3D sketching, moving beyond industrial design and including storytelling, medical applications as well as applications for cultural heritage, thereby demonstrating the versatility of sketch interactions.

10.1 Future Considerations

Although sketch interactions in both the 2D and 3D domains have improved considerably in terms of the flexibility afforded and the complexity of drawings supported by the sketching interface, there are open-issues that need to be addressed. Here we discuss some of these issues.

10.1.1 Combining Hand-Crafted Features and Deep-Learning

While traditional approaches to sketching simplification focus on the selection of features that correspond to the psycho-visual constructs that govern the human understanding of the sketch, in deep-learning approaches these constructs are learned through training. This has the advantage that the convolutional kernels learned are no longer restricted by the few selected constructs chosen for implementation, but can be learnt and adapted from the training examples provided.

A potentially beneficial avenue for exploration is to combine lessons learned from traditional approaches with the advantages of deep-learning by integrating within the CNN architectures layers that specifically incorporate the psycho-visual constructs that guide the human interpretation of the sketch, for example, the constructs that mimic the pattern grouping and inhibition associated with the visual cortex. Using architectures such as that proposed in Ayyoubzadeh and Wu [1], for example, where filter-bank regularisation is used to encourage the CNN kernels to conform to spatial structures and filters without overly constraining the CNN kernels.

10.1.2 Availability of Datasets

One important consideration in deep-learning approaches is the high dependency on the datasets available for training. Datasets collected from a small group of sketchers risk emphasizing the idiosyncrasies of the sketcher, with the consequence that learned models do not necessarily generalize well to other sketch types. This has driven researchers to curate datasets that reflect the diversity in sketching styles and varying degrees of roughness, and their corresponding ground-truth images. The dataset described by Yan et al. [20] is one such comprehensive dataset containing 2D sketches and their ground-truths both as simplified images and in vector format.

Nevertheless, obtaining a reasonably sized dataset with 3D objects that are rendered from the user sketches would be beneficial to have to ensure that the reconstruction algorithms can be evaluated with "in the wild" sketches.

10.1.3 Creating Full 3D Models from Sketches

The problem remains that of generating 3D models which are robust to drawing roughness, while at the same time retaining the shape expected from the sketch. This requires a compromise between surface smoothness and retaining sharp ridges and edges that may exist on the surface. Another issue, of course, remains that of determining a full 3D model from the 2D sketch since this requires some guess-work on what the unsketched part of the object looks like. Multiple views and interactive sketching on the surface of the 3D object, thereby bringing together the digital inking and 3D sketching techniques described in Chapter 3 and Chapter 6 respectively, would allow users to manipulate the sketched reconstruction until the result is visually pleasing and matches the user intent.

10.1.4 Sketch Simplification in the Presence of Shadows

An open question in the simplification process is the extraction and simplification of shadows or shading cues. These have an important role in the human perception of depth but, because of the diversity in the stylization used to represent shadows and shading strokes are not as widely used in drawing interpretation algorithms as they are in the interpretation of a scene imagery. Moreover, unlike scene imagery, sketched shadows and shading cues are sparse and prone to errors such that algorithms that use such cues for object interpretation must be robust to these additional difficulties. Some work in the integration of shadow cues has been proposed for line labeling algorithms, such as Bonnici and Camilleri [3], although much more can be done to determine surface characteristics through shading.

10.1.5 Long-Term Usability Studies

Although headsets and controllers are becoming lighter, ergonomics and user comfort remains an important consideration in the hardware design of immersive systems, particularly if one is to take into account the prolonged usage of these systems. It is also important to take into account user fatigue that can also stem from the use of immersive systems, particularly when taking into account issues such as the vergence-accommodation conflicts that may arise when the user needs to focus on the close screen portraying a far deeper object, or the sensory-induced sickness when the rendered object or sketch and the physical reality mismatch. Attention must, therefore, be given to rendering speeds and computational efficiency to minimize the lag between the virtual and physical worlds. This, in part, depends on the choice of algorithms for tracking, sampling, and rendering of the user strokes, but will also depend on the computational resources available for rendering the virtual environment in real-time, especially when rendering computationally-heavy lightning and shading effects. Rendering fully only the region in focus would minimize the computational costs but requires accurate tracking of the user's point of regard.

10.1.6 Dealing with Inaccurate Drawings

In the design of sketching interactions in the 3D domain, it is also important to keep in mind that sketching in 3D space may be inherently inaccurate due to the lack of the physical structure that guides the pen placement. Introducing devices such as scaffolding or haptic feedback among others would help the user maintains sketching accuracy. However, these measures, as well as the hand-held controllers and other immersive sketching techniques require time for acclimatization. Immersive systems described in the literature, however, only describe user studies conducted over short periods of time, and often with users already having some experience in immersive systems. This brings about the need for long-term usability studies with a broader range of users with different experiences from different fields, to assess the long-term effect on user interactions with the VR/AR system.

10.1.7 Accomodating Different Workflows

The different workflows in the design process give rise to different sketches, often by different designers working at different phases in the project. Thus, another important aspect of the design process is the ability to track through the different sketches, for example, to identify when a specific design decision was taken. The concept of product life-cycle management in product design is a management system that holds all information about the product, as it is produced throughout all phases of the product's life cycle. Such systems typically include product information in textual or organizational chart formats, providing information to different key actors along the product's life cycle on aspects such as recyclability and reuse of the different parts of the product, modes of use, and more [13]. This information is typically made available to different key members in an organization, from those at managerial and technical levels to key suppliers and customers [19]. Expanding this information with the sketches, drawings and 3D modeling information carried out during the design process will, therefore, extend the information contained in the document. Consumers would be able to trace back to the design decisions of particular features on the object, while designers would be able to understand how consumers are using the product and could exploit this information, for example, to improve quality goals. The challenge, therefore, lies in providing the means to establish an indexing and navigation system of the product design history, providing a storyboard of the design process from ideation stages to the final end-product.

One may note that, although a considerable number of sketching systems exist as research-based systems, few systems actually make their way into the actual work-place. We believe that this may be due to a number of factors. First of all, different workplaces utilize their own workflows and graphic communication systems and a complete overhaul of working systems to introduce newer, experimental systems is natural to meet resistance. It is, therefore, important when developing lab-based systems to keep in mind the possibility of integrating these novel systems with existing workflows, and this would include ensuring that the input and output file formats are compatible with the most commonly used systems, thus ensuring the continued usability of the novel interface outcome. Another point to consider is that different users have different preferences. For instance, while an individual may be open to viewing and working with prototypes in immersive VR/AR environments, their preferred method for conceptual sketching may be pen-and-paper. Thus, it becomes necessary to offer users a suite of tools with the ability to migrate between tools or start from any point in the workflow. This requires some handshaking between different interfaces, something which does not currently look into since many of the research-based systems are often investigated in isolation.

10.1.8 Multi User Collaboration

While user communication with the sketching interface is of paramount importance, one must not neglect communication between multiple users, specifically, the communication of nonverbal cues which enhance the collaborative experience between users. This communication too needs to be based on the natural interactions and gestures used by users, including for example, the focus of attention on the shared media at any given instance [6]. Such interactions would typically include eye-gaze and head-pose monitoring as well as finger-pointing and other hand gestures [17, 14, 4, 8]. The need for remote collaborative interactions is now felt more than ever. This needs to take into account that remote collaborators do not necessarily have accessibility to all tools available in laboratory environments. Thus, here too, the collaborative tools should be offered as a suite of tools that can be adapted starting from webcams which all users are most likely to have, to HMDs and other VR/AR systems. Care is required, however, since real-time visualizations based on these interactions can be distracting because of their low signal-to-noise ratios. Thus, further research is needed on the fusion of non-verbal cues together with speech information such that co-resolution of the point of regard of different users, particularly in multi group settings can be visualized effectively. One must also take into consideration that at conceptual design stages, the focus is on the quick generation and exploration of ideas, Thus, systems that support concept sketches should also focus on providing results quickly. It is also necessary to acknowledge that the results are not necessarily the finalized prototype and that the user may want to continue evolving, adjusting or correcting the prototype. Such continuous changes and adjustments should be easy to execute.

10.1.9 Accessibility of the Design Process

A further consideration that must be taken into account is the accessibility to these tools by people with disabilities. Unfortunately, the heavy usage of graphical user interfaces and particularly, the drive to have minimalist sketching interfaces that mimic the blank sheet of paper, are designed with sighted and able-bodied users in mind [12]. To prevent exclusion sketching interfaces and sketch-based modeling applications need to provide alternative tools to allow for sketching and visualisation of the rendered 3D models. For people with motor disabilities, this may require designing interfaces that could be used with control devices other than the mouse. This could be head movements [16], eye control [11], or voice control [10, 5], all of which allow users to draw using geometric shapes as well as freehand lines. The preferred input modality may well depend on the user's physical abilities and, even within any preferred modality, the individual user may have particular preferences such as locating buttons to be within the user's motor range, or adjusting the size of the buttons according to the degree of control managed by the user. Kwan and Betke [16] advocate for user interfaces that are highly customizable to meet the needs of the individual.

The interfaces described in Harada et al. [10], Hornof and Cavender [11], Kwan and Betke [16] still rely on visual feedback to show the user where the sketch is being drawn. Such feedback is not suitable for people with visual impairments. Blind and visually impaired persons require some form of haptic or audio feedback to help them locate their position to start drawing and to go back to the drawing for modification or continued drawing [12]. Completely offline raised-line drawing kits exist, allowing the user to place a piece of plastic sheet on a rubberized board, using a ball-point pen for drawing. The pressure of the pen will raise lines which will provide the user with haptic feedback [9]. For the intent of sketching interfaces, this drawing could then be processed as one would a pen-and-paper drawing. Haptic feedback for drawing may also be obtained through tactile drawing tablets which have Braille-like pins that rise when traced over [9]. Alternatively, rather than use haptic feedback, it is also possible to draw using voice commands coupled with a recursive grid system which allow users to create digital art based on line primitives by vocally selecting the starting and end positions using a numbered scheme that users are already familiar with [12]. Another possibility would be to allow users who prefer to use CAD programming in conjunction with existing screen readers to do so, hence bypassing the sketching part and going directly to create 3D models [18].

Blind and visually impaired users have to overcome the additional hurdle of visualizing the 3D designs. While it is possible to 3D print the designs for haptic feedback, this will be costly and time consuming if each modification to the prototype needs to be printed. Alternative haptic feedback options are necessary. Tactile, refreshable displays allow 2.5D rendering of the object being designed, allowing near-instantaneous update of the 3D model [18]. Virtual reality, specifically haptic feedback through wearable gloves will also allow the user to "feel" a virtual object [15]. Thus, while tools and alternatives exist to allow differently-abled users the chance to engage in design, existing sketching interfaces need to be open for the integration of these tools in the sketch-to-3D model pipeline.

10.2 Conclusion

To conclude, in this book, we presented the state of the art in sketch-based interactions, their limitations, and future directions. While the field has matured significantly and recent developments in deep-learning have extended the capabilities in the field beyond what was considered possible, more effort should be dedicated to integrating different sketch-interaction methodologies to provide a holistic, user-centric solution that moves beyond research labs and into practical work-places. We hope that this book provides the space for continued discussions in this direction.

References

[1] S. M. Ayyoubzadeh and X. Wu. Filter bank regularization of convolutional neural networks, 2019. arXiv preprint arXiv:1907.11110

[2] A. Bartolo, K. P. Camilleri, S. G. Fabri, J. C. Borg, and P. J. Farrugia. Scribbles To Vectors: Preparation Of Scribble Drawings For Cad Interpretation. In *Proceedings of the 4th Eurographics Workshop on Sketch-Based Interfaces and Modeling*, SBIM '07, page 123–130, 2007.

[3] A. Bonnici and K. P. Camilleri. A Constrained Genetic Algorithm For Line Labelling Of Line Drawings With Shadows And Table-lines. *Computers & Graphics*, 37(5):302 – 315, 2013.

[4] M. Cherubini, M. A. Nüssli, and P. Dillenbourg. Deixis and Gaze in Collaborative Work at a Distance (over a Shared Map): A Computational Model to Detect Misunderstandings. In *Proceedings of the 2008 Symposium on Eye Tracking Research & Applications*, ETRA '08, page 173–180, 2008.

[5] L. Dai, R. Goldman, A. Sears, and J. Lozier. Speech-Based Cursor Control: A Study of Grid-Based Solutions. In *Proceedings of the 6th International ACM SIGACCESS Conference on Computers and Accessibility*, Assets '04, page 94–101, 2003.

[6] S. D'Angelo and D. Gergle. An Eye For Design: Gaze Visualizations for Remote Collaborative Work. In *Proceedings of the 2018 CHI Conference on Human Factors in Computing Systems*, pages 1–12, 2018.

[7] L. Donati, S. Cesano, and A. Prati. A Complete Hand-drawn Sketch Vectorization Framework. *Multimed Tools and Applications*, 78: 19083–19113, 2019.

[8] J. Eisenstein, R. Barzilay, and R. Davis. Gesture Salience as a Hidden Variable for Coreference Resolution and Keyframe Extraction. *Journal in Artificial Intelligence Research*, 31(1):353–398, February 2008.

[9] T. Götzelmann. Visually Augmented Audio-Tactile Graphics for Visually Impaired People. *ACM Trans. Access. Comput.*, 11(2), 2018.

[10] S. Harada, J. O. Wobbrock, and J. A. Landay. Voicedraw: A hands-free voice-driven drawing application for people with motor impairments. Assets '07, page 27–34, 2007.

[11] A. J. Hornof and A. Cavender. EyeDraw: Enabling Children with Severe Motor Impairments to Draw with Their Eyes. In *Proceedings of the SIGCHI Conference on Human Factors in Computing Systems*, CHI '05, page 161–170, 2005.

[12] H. M. Kamel and J. A. Landay. Sketching Images Eyes-Free: A Grid-Based Dynamic Drawing Tool for the Blind. In *Proceedings of the Fifth International ACM Conference on Assistive Technologies*, Assets '02, page 33–40, 2002.

[13] D. Kiritsis. Closed-loop PLM For Intelligent Products In The Era Of The Internet Of Things. *Computer-Aided Design*, 43(5):479–501, 2011.

[14] D. Kirk, T. Rodden, and D. S. Fraser. Turn It This Way: Grounding Collaborative Action with Remote Gestures. In *Proceedings of the SIGCHI Conference on Human Factors in Computing Systems*, CHI '07, page 1039–1048, 2007.

[15] J. Kreimeier, P. Karg, and T. Götzelmann. Tabletop Virtual Haptics: Feasibility Study for the Exploration of 2.5D Virtual Objects by Blind and Visually Impaired with Consumer Data Gloves. In *Proceedings of the 13th ACM International Conference on Pervasive Technologies Related to Assistive Environments*, PETRA '20, 2020.

[16] C. Kwan and M. Betke. Camera Canvas: Image Editing Software for People with Disabilities. In *Proceedings of the 6th International Conference on Universal Access in Human-Computer Interaction: Applications and Services - Volume Part IV*, UAHCI'11, page 146–154, 2011.

[17] A. F. Monk and C. Gale. A Look Is Worth a Thousand Words: Full Gaze Awareness in Video-Mediated Conversation. *Discourse Processes*, 33 (3):257–278, 2002.

[18] A. F. Siu, S. Kim, J. A. Miele, and S. Follmer. ShapeCAD: An Accessible 3D Modelling Workflow for the Blind and Visually-Impaired Via 2.5D Shape Displays. In *The 21st International ACM SIGACCESS Conference on Computers and Accessibility*, ASSETS '19, page 342–354, 2019.

[19] R. Sudarsan, S. J. Fenves, R. D. Sriram, and F. Wang. A Product Information Modeling Framework For Product Lifecycle Management. *Computer-Aided Design*, 37(13):1399–1411, 2005.

[20] C. Yan, D. Vanderhaeghe, and Y. Gingold. A Benchmark for Rough Sketch Cleanup. *ACM Transactions on Graphics*, 39(6), November 2020.

Index

3D mesh 201, 204, 243, 251

About the Editors

Professor Kenneth P. Camilleri graduated with a B.Elec.Eng.(Hons) degree in electrical engineering from the University of Malta in 1991, and with an M.Sc. in Signal Processing and Machine Intelligence and Ph.D. degrees from the University of Surrey, UK in 1994 and 1999, respectively.

Professor Camilleri's research work is in the fields of signal processing, computer vision and machine learning. His specific research interests include automatic drawing interpretation, eye gaze tracking, human computer interfacing based on electrophysiological signals such as the electroencephalogram (EEG) and the electrooculogram (EOG), and biomedical signal and image analysis for physiological measurement and the extraction of disease biomarkers. He has published research in these areas in over 160 international peer-reviewed publications.

Professor Camilleri is a founding member of the Department of Systems and Control Engineering, and of the Centre for Biomedical Cybernetics, both at the University of Malta. He presently serves as Director of the Centre for Biomedical Cybernetics. Professor Camilleri is a Senior Member of the Institute of Electronic and Electrical Engineers (IEEE, USA), the Society for Photo-Optical Instrumentation Engineering (SPIE, USA), and the Association for Computing Machinery (ACM, USA), and a member of the Institution of Engineering and Technology (IET, UK).

Dr Alexandra Bonnici received her B.Eng (Hons) in 2004, an MPhil (Melit.) in Engineering in 2008 and PhD (Melit.) in 2015 on work related to the vectorization and interpretation of sketches with artistic cues.

Alexandra's research work is the fields of computer vision and machine learning. Specific research interests include the machine interpretation of drawings and sketches, text and music document processing as well as the development of applications which assist the teaching and learning of music.

Alexandra is a Senior Lecturer at the University of Malta and is currently serving as Head of Department of the Department of Systems and Control Engineering. She is also the coordinator of the Faculty's Certificate in Engineering Sciences. She is also active in promoting STEM education, having set up the Engineering Technology Clubs at the Faculty of Engineering. Alexandra is an Senior Member of the Institute of Electronic and Electrical Engineers (IEEE, USA), a member of the Association for Computing Machinery (ACM, USA) and the Eurographics Association. She is also a steering committee member of the ACM Symposium on Document Engineering.